Two Manuals Combined into One Book

Combat Breaching

and

Combat Engineer

Department of Defense

Pentagon
Publishing

Field Manual
No. 3-34.2

Headquarters
Department of the Army
Washington, DC, [pending date]

Combined-Arms Breaching Operations

Contents

Distribution Restriction: Approved for public release; distribution is unlimited.

*This publication supersedes Field Manuals (FMs) 5-101, 23 January 1985, and 90-13-1, 28 February 1991.

Page

Page

Preface

FM 3-34.2 provides combined-arms commanders, from company to brigade, and their staffs with the doctrine, tactics, and techniques needed to successfully overcome obstacles. It provides the commander with breaching fundamentals and information for planning, preparing, and executing breaching operations. This manual gives the coordinating and special staff officer a basis for synchronizing all combat multipliers during a breaching operation.

This is not a stand-alone manual. The user must have a fundamental understanding of the concepts outlined in FMs 5-71-2, 5-71-3, 5-71-100, 17-98, 20-32, 34-130, 71-1, 71-2, 71-3, 100-5, 100-7, 101-5, and 101-5-1. This manual also implements Standardization Agreement (STANAG) 2036.

Appendix A contains a metric conversion chart.

The proponent for this publication is HQ, TRADOC. Send comments and recommendations on Department of the Army (DA) Form 2028 directly to Commandant, United States (US) Army Engineer School, ATTN: ATSE-DOT-DD, Directorate of Training, 320 Engineer Loop, Suite 336, Fort Leonard Wood, Missouri 65473-8929.

Unless this publication states otherwise, masculine nouns and pronouns do not refer exclusively to men.

Chapter 1

Obstacle Breaching Theory

Breaching operations are conducted to allow maneuver despite the presence of obstacles. Obstacle breaching is the employment of a combination of tactics and techniques to advance an attacking force to the far side of an obstacle that is covered by fire. It is perhaps the single, most difficult combat task a force can encounter. Understanding breaching theory is the first step to understanding breaching tactics. Breaching is a synchronized combined-arms operation under the control of a maneuver commander. Breaching operations begin when friendly forces detect an obstacle and begin to apply the breaching fundamentals, and they end when battle handover has occurred between follow-on forces and a unit conducting the breaching operation. *Bulling through* or *forcing through* is not a breaching operation. Bulling through is a decision made when a commander must react immediately to extricate his force from an untenable position within an obstacle and no other breaching operations are possible. When a force is in a minefield receiving fires and taking heavy losses, the commander may decide to immediately bull through the minefield rather than withdraw or reduce the obstacle.

DEFINITIONS

1-1. The following definitions are inherent to obstacle breaching and are used throughout this manual:

- **Obstacle**. An obstacle is any obstruction that is designed or employed to disrupt, fix, turn, or block the movement of an opposing force (OPFOR) and to impose additional losses in personnel, time, and equipment on the OPFOR. Obstacles can exist naturally (existing), be man-made (reinforcing), or be a combination of both. A complex obstacle is a combination of different types of individual obstacles that requires more than one reduction technique (explosive, mechanical, manual) to create a lane through the obstacle. A reinforcing obstacle is an obstacle that is specifically constructed, emplaced, or detonated through military effort. (For more information on obstacle classification, see *FM 90-7*.) The types of reinforcing obstacles are—
  - **Tactical**. A tactical obstacle is employed to disrupt enemy formations, turn them into a desired area, fix them in position under direct and indirect fires, or block their penetration while multiplying the effects and capabilities of firepower.
  - **Protective**. A protective obstacle is employed to assist a unit in its local, close-in protection.

- **Lane**. A lane is a route through, over, or around an enemy or friendly obstacle that provides safe passage of a passing force. The route may be reduced and proofed as part of a breaching operation, be constructed as part of the obstacle, or be marked as a bypass.
- **Reduction**. This is a task to create and mark lanes through, over, or around an obstacle to allow the attacking force to accomplish its mission. Normally, engineers and reduction assets are used to reduce an obstacle.
- **Proofing**. Proofing verifies that a lane is free of mines and that the width and trafficability of the point of breach are suitable for the assault force. Proofing can be conducted visually (against surface-laid minefields), electronically (mine detectors), or mechanically (mine-clearing rollers [MCRs]). Proofing is conducted when the risk of live mines remaining in the lane exceeds the risk of loss (lives and equipment) to enemy fires while waiting to complete proofing. Some mines are resistant to some reduction assets; for example, magnetically fused mines may be resistant to a mine-clearing line charge (MICLIC). Proofing is vitally important, considering the vast variety of mines in use. It should always be planned for a breaching operation; however, the time available, the threat, or the mission may dictate that proofing not be done.
- **Clearing**. This is the total elimination or neutralization of an obstacle or a portion of the obstacle. Clearing operations are not conducted under fire and are usually performed by follow-on engineer forces.
- **Breach area**. The breach area is established and fully defined by the higher headquarters (HQ) of the unit conducting breaching operations. It is the area where a breaching operation occurs. The area must be large enough to allow the attacking unit to deploy its support force and extend far enough on the far side of the obstacle to allow follow-on forces to deploy before leaving the breach area. One technique is to establish the breach area using phase lines (PLs) or unit boundaries. The PL defining the far side of the breach area may be established as a battle handover line (BHL).
- **Farside objective**. The farside objective is the immediate objective of the attacking unit's assault force, and it may be oriented by the terrain or the enemy force. The higher HQ assigns the objective; however, the attacking unit normally subdivides the objective into smaller objectives to assign responsibilities and to control and focus the assault of subordinate forces. When breaching as part of a larger force, seizing the farside objective provides the necessary maneuver space for the higher unit's follow-on forces to move securely through the lanes, assemble or deploy, and continue the attack without enemy interference.
- **Bypass**. A bypass is a tactical task that involves maneuvering around an obstacle, a position, or an enemy force to maintain the momentum of advance. When a unit bypasses an obstacle, it physically changes the direction of movement to avoid the obstacle. This must be done with caution because it might play into the enemy's hand. Bypassed obstacles and enemy forces are reported to higher HQ.

- **Point of penetration**. This is the location, which is identified on the ground, where the commander concentrates his efforts at the enemy's weakest point to seize a foothold on the farside objective. This is achieved along a narrow front through maneuver and direct and indirect fires that are accurately placed against enemy forces. A commander conducting a breaching operation establishes a point of penetration that supports planning locations for the reduction area and the seizure of the farside objective.
- **Reduction area**. This is a number of adjacent points of breach that are under the control of the breaching commander. The commander conducting the attack determines the size and location of the reduction area that supports the seizure of a point of penetration (see *Figure 1-1*).
- **Point of breach**. This is the location at an obstacle where the creation of a lane is being attempted. Initially, points of breach are planned locations only. Normally, the breach force determines the actual points of breach during the breaching operation.

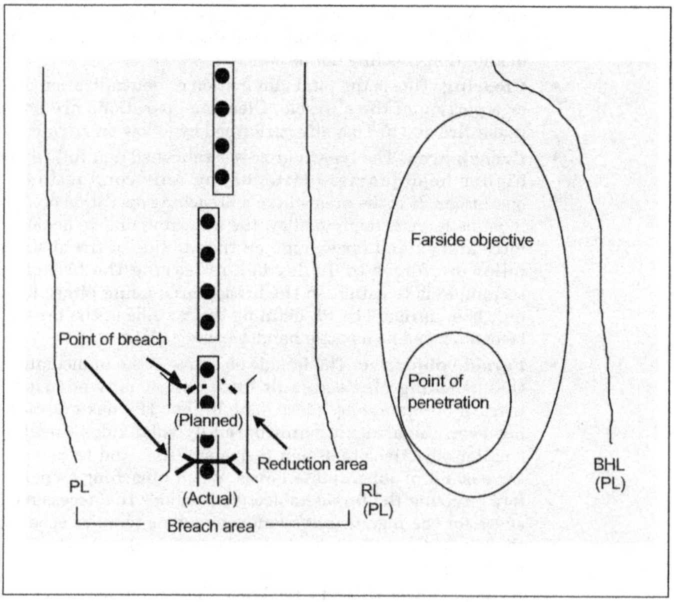

Figure 1-1. Breach Area

BREACHING TENETS

1-2. Successful breaching operations are characterized by applying breaching tenets. These tenets should be applied whenever an obstacle is encountered in the area of operations (AO), whether during an attack or during a route clearance operation. The tenets are—

- Intelligence.
- Breaching fundamentals.
- Breaching organization.
- Mass.
- Synchronization.

INTELLIGENCE

1-3. Success depends largely on the force commander's ability to *see the battle space*. He must identify how the enemy is using the terrain to minimize the risk of surprise. This is particularly true when attempting to counter the enemy's use of obstacles. The force commander does this with the intelligence preparation of the battlefield (IPB) process. During the IPB process, the situation template (SITEMP) is developed. The SITEMP is a graphic depiction of expected threat dispositions based on threat doctrine and the effects of the battle space for a particular course of action (COA). Based on the commander's estimates, the staff develops a SITEMP for multiple enemy COAs. The SITEMP helps drive the maneuver planning in support of war gaming and the development of the event template. The SITEMP also drives the decisions the commander must be prepared to make. The commander uses the decision-support template (DST) to convey these decisions in graphic form. The IPB process helps achieve success by focusing available intelligence-gathering assets toward answering the commander's priority intelligence requirements (PIR). Intelligence that is gathered later by recon becomes the foundation for developing and revising a SITEMP. (See *Appendix B* for information on threat mine operations.)

1-4. Time-constrained planning requires the rapid development of the enemy's SITEMP. A minimal SITEMP, complete with templated obstacles, allows the development of an initial event template and provides initial PIR for the intelligence-collection plan (ICP). At a minimum, the unit can template an obstacle trace that is precise, as far as the location, and represents the total expected enemy countermobility effort but does not depict individual obstacles. The staff plots the actual composition of the trace based on incoming obstacle reports. Adequate time for intelligence collection is critical in developing an accurate picture of the battle space, yet time is normally in short supply. While templating is essential to focus the ICP, too much time can be spent on templating at the expense of intelligence, surveillance, and reconnaissance (ISR).

1-5. In any operation where enemy obstacles can interfere with friendly maneuver, obstacle intelligence (OBSTINTEL) is always one of the information requirements (IR) and should become PIR. Finding enemy obstacles or seeing enemy obstacle activity validates and refines the Intelligence Officer's (US Army) (S2's) picture of the battle space.

OBSTINTEL helps the S2 determine the enemy's intentions, plans, and defense strength. The force engineer is the unit's expert on enemy countermobility capabilities. He assists the S2 in templating enemy obstacles, analyzing OBSTINTEL, and analyzing the effects of terrain.

1-6. An unverified enemy template can lead to disaster because the force may aim an attack at the wrong place. Units may deploy to reduce expected obstacles early, wasting mission time to *feel* their way into nonexistent obstacles; or they may blunder into an unexpected obstacle or an enemy engagement area (EA).

1-7. OBSTINTEL can be a critical indicator to verify an enemy template. The force's Operations and Training Officer (US Army) (S3), S2, and engineer establish effective OBSTINTEL collection by determining specific obstacle IR. Examples of information that is needed to fulfill obstacle IR include—

- The location of existing and reinforcing obstacles.
- The orientation and depth of obstacles.
- Conditions of the soil (in the case of a minefield) to determine the ability to use tank plows.
- The presence, location, and type of wire.
- Lanes and bypasses.
- The composition of the minefield (buried or surface-laid antitank [AT] and antipersonnel [AP] mines, antihandling devices [AHDs], and the depth of the mines).
- Types of mines and fuses.
- The location of enemy indirect-fire systems that can fire into the breach area.
- The composition of complex obstacles.
- Areas between successive obstacle belts.

1-8. OBSTINTEL is particularly important for discovering the types of mines and mine fuses the enemy has employed. The engineer depends on this information since he must determine which reduction techniques offer the best chance for success and minimize the risk to the breach force. This may require a dismounted recon patrol to examine mines within a minefield. The enemy may employ security patrols, have forces overwatching the obstacle, or place forces within the obstacle to prevent obstacle recon.

1-9. Obtaining OBSTINTEL requires dedicated collection assets. These assets identify survivability positions and obstacle emplacement activity. Aviation units, who are tasked to perform recon in the area, may be able to provide valuable OBSTINTEL information. Ground-surveillance radar may also provide useful information.

1-10. Recon forces should have the capability to collect OBSTINTEL. To be successful in this collection, the obstacle recon efforts must be tied into the maneuver force's overall reconnaissance and surveillance (R&S) plan. A force ensures this capability by training dedicated recon assets on collecting required obstacle information or by augmenting recon forces with engineer recon teams provided by the supporting engineer force. An engineer recon team moves with scouts or a patrol and conducts dismounted recon of templated or confirmed obstacles. This team has the flexibility to dismount and develop the details of the obstacle. Personnel engaged in a recon mission for OBSTINTEL should be used rarely, if ever, to reduce obstacles during the mission; although they make ideal guides to breach forces for subsequent breaching operations. Inadvertent detonation during reduction may compromise engineers and scouts, defeating the recon mission. It may also compromise the entire attack. *FM 5-170* describes capabilities, limitations, and employment concepts of engineer recon teams.

BREACHING FUNDAMENTALS

1-11. Suppress, obscure, secure, reduce, and assault (SOSRA) are the breaching fundamentals that must be applied to ensure success when breaching against a defending enemy. These fundamentals will always apply, but they may vary based on the specific battle-space situation (mission, enemy, terrain, troops, time available, and civilian considerations [METT-TC]).

Suppress

1-12. Suppression is a tactical task used to employ direct or indirect fires or an electronic attack on enemy personnel, weapons, or equipment to prevent or degrade enemy fires and observation of friendly forces. The purpose of suppression during breaching operations is to protect forces reducing and maneuvering through an obstacle.

1-13. Effective suppression is a mission-critical task performed during any breaching operation. Suppressive fires in sufficient volume serve to secure the reduction area. Successful suppression generally triggers the rest of the actions at the obstacle. Fire control measures ensure that all fires are synchronized with other actions at the obstacle. Although suppressing the enemy overwatching the obstacle is the mission of the support force, the breach force should be able to provide additional suppression against an enemy that the support force cannot effectively suppress.

Obscure

1-14. Obscuration must be employed to protect forces conducting obstacle reduction and the passage of assault forces. Obscuration hampers enemy observation and target acquisition and conceals friendly activities and movement. Obscuration smoke deployed on or near the enemy's position minimizes its vision. Screening smoke employed between the reduction area and the enemy conceals movement and reduction activities. It also degrades enemy ground and aerial observations. Obscuration must be carefully planned to provide maximum degradation of enemy observation and fires, but it must not significantly degrade friendly fires and control.

Secure

1-15. Friendly forces secure the reduction area to prevent the enemy from interfering with obstacle reduction and the passage of the assault force through the lanes created during the reduction. Security must be effective against outposts and fighting positions near the obstacle and against overwatching units, as necessary. The far side of the obstacle must be secured by fires or be occupied before attempting any effort to reduce the obstacle. The attacking unit's higher HQ has the responsibility to isolate the breach area by fixing adjacent units, attacking enemy reserves in depth, and providing counterfire support.

1-16. Identifying the extent of the enemy's defenses is critical before selecting the appropriate technique to secure the point of breach. If the enemy controls the point of breach and cannot be adequately suppressed, the force must secure the point of breach before it can reduce the obstacle.

1-17. The breach force must be resourced with enough maneuver assets to provide local security against the forces that the support force cannot sufficiently engage. Elements within the breach force that secure the reduction area may also be used to suppress the enemy once reduction is complete.

Reduce

1-18. Reduction is the creation of lanes through or over an obstacle to allow an attacking force to pass. The number and width of lanes created varies with the enemy situation, the assault force's size and composition, and the scheme of maneuver. The lanes must allow the assault force to rapidly pass through the obstacle. The breach force will reduce, proof (if required), mark, and report lane locations and the lane-marking method to higher HQ. Follow-on units will further reduce or clear the obstacle when required. Reduction cannot be accomplished until effective suppression and obscuration are in place, the obstacle has been identified, and the point of breach is secure. Obstacle-reduction techniques are discussed in more detail in *Appendix C*.

Assault

1-19. A breaching operation is not complete until—

- Friendly forces have assaulted to destroy the enemy on the far side of the obstacle that is capable of placing or observing direct and indirect fires on the reduction area.
- Battle handover with follow-on forces has occurred, unless no battle handover is planned.

BREACHING ORGANIZATION

1-20. A commander organizes friendly forces to accomplish the breaching fundamentals quickly and effectively. This requires him to organize support, breach, and assault forces with the necessary assets to accomplish their roles (see *Table 1-1*, page 1-8).

Table 1-1. Relationship Between Breaching Organization and Breaching Fundamentals

Breaching Organization	Breaching Fundamentals	Responsibilities
Support force	Suppress Obscure	Suppress enemy direct-fire systems covering the reduction area. Control obscuring smoke. Prevent enemy forces from repositioning or counterattacking to place direct fires on the breach force.
Breach force	Suppress (provides additional suppression) Obscure (provides additional obscuration in the reduction area) Secure (provides local security) Reduce	Create and mark the necessary lanes in an obstacle. Secure the nearside and far side of an obstacle. Defeat forces that can place immediate direct fires on the reduction area. Report the lane status/location.
Assault force	Assault Suppress (if necessary)	Destroy the enemy on the far side of an obstacle that is capable of placing direct fires on the reduction area. Assist the support force with suppression if the enemy is not effectively suppressed. Be prepared to breach follow-on and/or protective obstacles after passing through the reduction area.

Support Force

1-21. The support force's primary responsibility is to eliminate the enemy's ability to interfere with a breaching operation. It must—

- Isolate the reduction area with fires.
- Mass and control direct and indirect fires to suppress the enemy and to neutralize any weapons that are able to bring fires on the breach force.
- Control obscuring smoke to prevent enemy-observed direct and indirect fires.

1-22. Suppression is critical for a successful breaching operation; therefore, allocation of adequate forces to the support force is essential. The commander must allocate enough direct- and indirect-fire systems to the support force to allow it to accomplish its missions. When resourcing the support force, consider the losses it will receive as it fights its way into its support-by-fire (SBF) position. To increase the survivability of the support force, the commander may establish critical friendly zones (CFZs) in support of the support force once it has occupied its SBF positions. A CFZ is an area, usually a friendly unit or location, that the maneuver commander designates as critical to protect an asset whose loss would seriously jeopardize the mission. Covered by a radar sector, the CFZ supports counterfire operations by providing the most responsive submission of targets to the fire-support system when rounds impact inside the CFZ. (For more detailed information on CFZs, see *FM 6-121*.)

1-23. The support force must seize an SBF position, seeking maximum protection from the terrain. If possible, the support force should follow a covered or concealed route to the SBF position, take up its assigned sectors of fire and observation, and begin to engage the enemy. It may have to make some adjustments to its direct-fire plan if the situation dictates. Observation is particularly critical. Artillery observers with the support force may initially bring indirect fires on enemy positions to fix and suppress the enemy. The support force adjusts the artillery-delivered obscuring smoke, if necessary, to protect the breach and assault forces as they approach the reduction area.

Breach Force

1-24. The breach force assists in the passage of the assault force by creating, proofing (if necessary), and marking lanes. (For more information on the initial lane-marking pattern, see *Chapter 3*.) The breach force is a combined-arms force. It includes engineers, reduction assets, and enough maneuver forces to provide additional suppression and local security. The breach force may apply portions of the following breaching fundamentals as it reduces an obstacle:

- **Suppress**. The breach force must be allocated enough maneuver forces to provide additional suppression against various threats, including—
  - Enemy direct-fire systems that cannot be effectively observed and suppressed by the support force due to the terrain or the masking of the support force's fires by the breach force as it moves forward to reduce the obstacle.
  - Counterattacking and/or repositioning forces that cannot be engaged by the support force.
- **Obscure**. The breach force employs vehicle-mounted smoke systems and smoke pots, if necessary, for self-defense and to cover lanes while the assault force is passing.
- **Secure**. The breach force secures itself from dismounted threat forces who are providing close-in protection of the obstacle. The breach force also secures the lanes through the tactical obstacles, once they are created, to allow safe passage of the assault force.
- **Reduce**. The breach force performs its primary mission of reducing the obstacle. To support the development of a plan to reduce the obstacle, the composition of the obstacle system must be an IR.
- **Assault**. The breach force assaults through the point of breach to the far side of an obstacle and seizes the reduction area.

1-25. The breach force is comprised of two subordinate elements—security and reduction:

- **Security element**. The security element is comprised primarily of maneuver forces. It is normally responsible for providing additional suppression, obscuration, and local security.
- **Reduction element**. The reduction element is comprised primarily of reduction assets (mine-clearing blades [MCBs], MICLICs, armored vehicle-launched bridges [AVLBs], armored combat earthmovers

[ACEs], dismounted engineers, and so forth). It is normally responsible for reducing the obstacle.

Chapter 2 provides more details on the roles, responsibilities, and command and control (C^2) of the security and reduction elements.

1-26. The breach force must be able to deploy and begin reducing the obstacle as soon as enemy fires have been suppressed and effective obscuration is achieved. It can expect enemy artillery fires within a matter of minutes. CFZs should be activated at the point of breach before the commitment of the breach force.

1-27. Engineers within the breach force are allocated with the reduction assets necessary to clear mines, nonexplosive obstacles, and small gaps. Scouts, engineers, or other recon assets are also capable of finding local bypasses or existing lanes in the obstacle system.

1-28. The commander allocates engineers and reduction assets based on the number of lanes required, the terrain, and the type of obstacle. The breach force must be able to create a minimum of one lane for each assaulting company or two lanes for the assaulting task force (TF).

1-29. The commander should plan for a 50 percent loss of mobility assets in close combat. As a planning factor, he should allocate one engineer platoon with the necessary reduction assets to create and mark one lane.

1-30. After the breach force has reduced the obstacle and passed the assault force through, the breach force may be required to hand over the lane to follow-on units. At a minimum, the lanes must be marked and their locations and conditions reported to higher HQ and follow-on units as prescribed in the unit's standing operating procedures (SOP).

Assault Force

1-31. The assault force's primary mission is to destroy the enemy and seize terrain on the far side of the obstacle to prevent the enemy from placing direct fires on the created lanes. The assault force may be tasked to assist the support force with suppression while the breach force reduces the obstacle.

1-32. The assault force must be sufficient in size to seize the point of penetration. Combat power is allocated to the assault force to achieve a minimum 3:1 ratio on the point of penetration. In a breaching operation, particularly TF and above, the assault force normally maneuvers as a separate force attacking through a reduced obstacle. However, breach and assault assets may maneuver as a single force when conducting lower-level breaching operations, such as an independent company team conducting an attack.

1-33. If the obstacle is defended by a small enemy force, assault and breach forces' missions may be combined. This simplifies C^2 and provides more immediate combat power for security and suppression.

1-34. Fire control measures are essential, since support and breach forces may be firing on the enemy when the assault force is committed. Suppression of overwatching enemy positions must continue and other enemy forces must remain fixed by fires until the enemy has been destroyed. The assault force must assume control for direct fires on the assault objective as support and breach forces' fires are lifted or shifted.

MASS

1-35. Breaching is conducted by rapidly applying concentrated efforts at a point to reduce the obstacle and penetrate the defense. Massed combat power is directed against the enemy's weakness. The location selected for breaching depends largely on the weakness in the enemy's defense, where its covering fires are minimized. If friendly forces cannot find a natural weakness, they create one by fixing the majority of the enemy force and isolating a small portion of it for attack.

Support Force

1-36. The support force suppresses the enemy to eliminate effective fires on the breach force. Smoke and terrain are used to help isolate the force under attack. Suppression requires the commander to mass enough overwatching direct fires to protect the breach force.

1-37. Normally, the TF isolates and destroys platoon-size elements in succession, starting with the platoon identified as the easiest to overwhelm. In *Figure 1-2*, the TF commander employs a tank company team to fix two defending platoons and uses screening smoke (target AA1102) to help isolate the platoon selected for attack. Objective (OBJ) Cowboy has been identified by the TF commander as the point of penetration. The center and eastern platoons are suppressed by direct and indirect fires from a second tank company team.

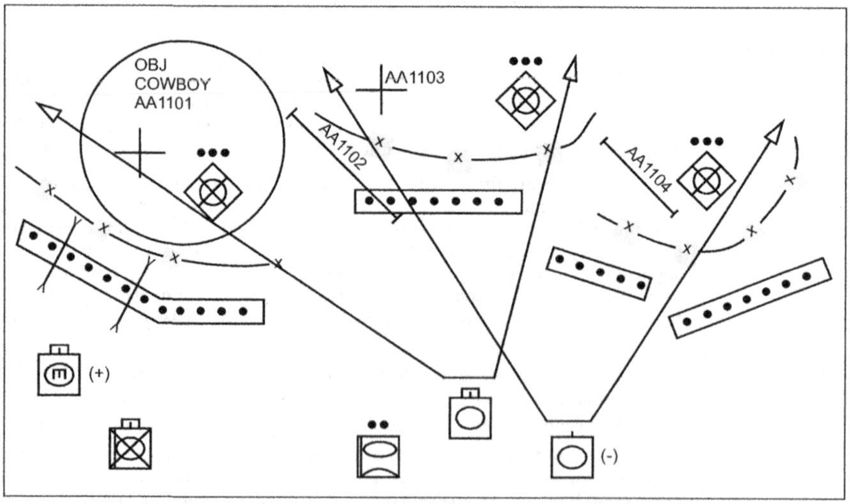

Figure 1-2. Technique for Achieving Mass in the Breach

1-38. The support force should be provided with reduction assets to reduce the impact that unexpected obstacles or scatterable minefields have on their approach to and occupation of SBF positions. Failure to do so may greatly impact on synchronizing the entire breaching operation. As a technique, a unit may create a mobility reserve that supports the main effort throughout the operation. Initially, the mobility reserve may support the support force until it has seized its SBF positions and then shift its support to the breach or assault force.

Breach Force

1-39. The commander also masses engineers and reduction assets to reduce the obstacle. The breach force is organized and equipped to use several different reduction techniques in case the primary technique fails; for example, if a key vehicle is destroyed or if casualties render dismounted engineers ineffective. Additional reduction assets are present to handle the unexpected. Normally, 50 percent more reduction assets than required for obstacle reduction are positioned with the breach force.

1-40. The breach force masses reduction efforts against the obstacle to ensure that it will successfully create enough lanes. A mounted TF-size assault force requires at least two lanes, but more lanes will speed passage through the danger area. The breach force attempts to simultaneously create as many lanes as necessary to ensure the rapid passage of the assault force through the obstacle system. This may require the breach force to simultaneously use a combination of mechanical, explosive, and manual techniques. For example, an engineer platoon may use two MICLICs as the primary technique to reduce a minefield. As soon as the engineer platoon fires the second MICLIC, engineer squads employ manual explosive techniques to create three additional lanes through the minefield. The commander must carefully plan and synchronize the creation of additional lanes to reduce the potential for fratricide with assaulting troops. The distance between the lanes depends on the enemy, the terrain, the need to minimize the effects of enemy artillery, the direct-fire plan of the support force, C^2, and reduction-site congestion.

Assault Force

1-41. Achieving necessary mass for the assault requires the breach force to open enough lanes through the obstacle to permit rapid passage and the buildup of forces on the farside. The size of the assault force determines the number of lanes initially created. A mounted TF requires one lane per assaulting company team. A dismounted assault force requires one lane for each leading assault platoon. The tactical situation may require additional lanes to quickly pass a large assault force through the obstacle to achieve sufficient combat-power ratio.

1-42. The principle of mass influences the selection of the point of breach; the task organization of the support, breach, and assault forces; and the integration of engineers in movement or attack formations. *Figure 1-2*, page 1-11, shows a technique to achieve mass in the breach.

1-43. The need to generate enough mass and the maneuver space available strongly influence which echelon can conduct the breaching operation. A company team cannot simultaneously mass sufficient fires, reduce the obstacle, and assault the defending position unless it is a simple obstacle defended by no more than one platoon. A TF should have a combat-power ratio of 3:1 to reduce an obstacle defended by a company, and a TF is normally the echelon used to execute a breach. A brigade has sufficient combat power to attack a complex, well-defended obstacle; but a brigade has difficulty deploying all its combat power within range. Normally, a brigade breaches by isolating a small segment of the defense (platoon or company) that the TF can then attack as the breaching echelon. If obstacles and defenses are in-depth, mass is achieved by passing additional TFs through the obstacle to continue the attack.

SYNCHRONIZATION

1-44. Breaching operations require precise synchronization of the breaching fundamentals by support, breach, and assault forces. Failure to synchronize effective suppression and obscuration with obstacle reduction and assault can result in rapid, devastating losses of friendly troops in the obstacle or the enemy's EA.

1-45. A combined-arms breach is a complex operation by nature. Support, breach, and assault forces must apply the SOSRA breaching fundamentals within a short time and distance. The support force masses its direct fires and controls indirect fires in concert with breach and assault forces' maneuvers. The commander must employ smoke at the right time and place to maximize its effectiveness or risk hampering his own target acquisition and C^2. The breach force must have the right reduction asset for the type of obstacle encountered. Engineers must ensure that they do not prematurely exhaust reduction assets needed to reduce subsequent obstacles. *Table 1-2*, page 1-14, shows the complexity of a combined-arms breach.

1-46. The commander ensures synchronization through proper planning and force preparation. Fundamentals to achieve synchronization are—

- Detailed reverse planning.
- Clear subunit instructions.
- Effective C^2.
- Well-rehearsed forces.

Detailed Reverse Planning

1-47. Synchronizing a combined-arms breach begins by using the reverse-planning process to ensure that actions at obstacles support actions on the objective. Planning the breach without regard to actions on the objective leads to disaster. During COA development, the staff analyzes the relative combat power and compares enemy and friendly strengths and weaknesses. The staff then generates options based on the commander's guidance and arrays the forces necessary to accomplish the mission. The staff develops the options by analyzing the objective and allocating forces to accomplish the mission. The commander decides how he must attack the objective to accomplish his mission. This decision drives where, how, and with what force he must support, breach, and assault through the enemy's obstacles. The commander approves the scheme of maneuver for the breaching phase of the operation

Table 1-2. Breach Complexity

Action	Element	Time (Minutes)	Controlled By
Develop the situation (verify the boundary of the enemy obstacle system).	Force in contact	M to 2	S3
Maneuver the support force into the overwatch position.	Support	M + 2 to 15	Support cdr
Maneuver the assault force into the covered assault position.	Assault	M + 2 to 15	Assault cdr
Call for artillery.	DS artillery	M + 2 to 15	FSO
Build smoke.	Mortars	M + 5 to 10	FSO
Suppress the enemy with direct fires.	Support	M + 15 to 29	Support cdr
Suppress the enemy with artillery fires.	DS artillery	M + 10 to 29	FSO
Maintain smoke.	DS artillery/mortars	M + 10 to 30	FSO
Maneuver the breach force to the breach location.	Breach	M + 20 to 23	Reduction cdr
Reduce the obstacle, and prepare two lanes.	Breach	M + 23 to 30	Engineer ldr
Place smoke pots.	Breach	M + 23 to EOM	Reduction cdr
Shift direct fires off the objective.	Support	M + 29 to 30	Assault cdr
Shift indirect fires beyond the objective.	DS artillery	M + 29 to 30	Assault cdr
Assault to destroy the enemy on the far side of the obstacle.	Assault	M + 30 to 45	Assault cdr
Reorganize to continue the mission.	TF	M + 45 to EOM	S3
NOTE: M = Contact with the obstacle			

that achieves adequate suppression, obscuration, and security. He also creates sufficient lanes to rapidly project combat power on the point of penetration and the far side of the objective, not just to the far side of the obstacle. Reverse planning gives purpose to the breaching plan that supports mission accomplishment.

1-48. The commander begins developing COAs as discussed in *FM 101-5*. He determines relative combat-power ratios, generates options, and arrays forces to accomplish the mission. He then develops the scheme of maneuver, including actions on the objective. In *Figure 1-3*, a three-company-team TF is conducting a deliberate attack to destroy a motorized infantry company (MIC) in the defense. The commander believes seizing OBJ Cat to be decisive. This is the point from which reverse planning of actions on the objective begins. The commander uses Team C to seize the initial foothold (OBJ Bird) (point of penetration) into the enemy's defense. The focus of breach planning is to get Team C to its objective. The number and location of lanes are driven by Team C's maneuver on OBJ Bird as the assault force. The commander now plans how the breach force must maneuver to reduce the obstacle. Supporting fires from Company A and Team B in SBFs 15 and 25 are planned to support the reduction of the obstacle and the assault on the objective. Reverse planning continues to drive the maneuver formation to ensure that forces are in the correct relative positions to accomplish their missions.

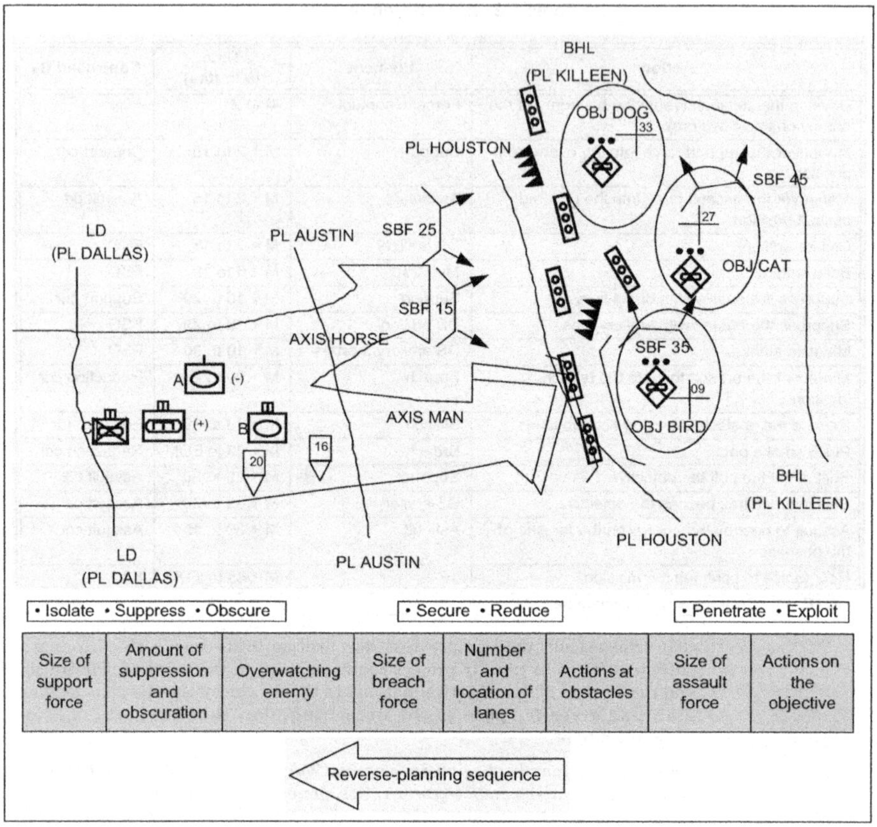

Figure 1-3. Reverse Planning

Clear Subunit Instructions

1-49. Subordinate units must clearly understand their missions, how their missions relate to the phases of the operation, and what role adjacent units play in the overall plan. The commander may use an execution matrix to synchronize his forces. An execution matrix is a superb synchronization tool, and it lists subunit instructions sequentially in relation to key events or the sequence of the attack. It also provides subordinate commanders with an understanding of how their missions and those of adjacent units fit into the overall plan. More importantly, it allows subordinates to better track the battle and coordinate their own maneuver with that of adjacent units. This is critical to achieving unity of effort between support, breach, and assault forces. *Figure 1-4* shows a sample execution matrix for the scheme of maneuver in *Figure 1-3,* page 1-15.

	Company A (Support Force)	**Team B (Support/ Assault Force)**	**Team C (Assault Force)**	**Team Engineer (Breach Force)**	
Units	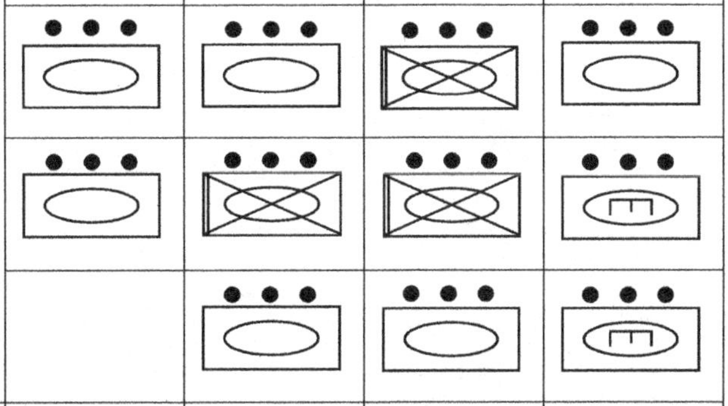				Force Allocation
LD to PL Austin	Move to left of Team B; echelon left.	Lead TF.	Trail team engineer.	Trail Team B center.	Maneuver formation
PL Austin to PL Houston	Attack with Team B on Axis Horse. Occupy SBF 25.	Lead attack on Axis Horse. Occupy SBF 15.	Occupy CP 20.	Lead attack on Axis Man.	
Obstacle reduction	SBF 25. Orient TRP 09. On order TRP 27.	SBF 15. Orient TRP 09.	Occupy CP 16. Identify lane.	Create and mark two lanes (vicinity AB123456).	
Attack on OBJ Bird	Shift fires to TRP 33.	Shift fires to TRP 27.	On order, move through lane to seize OBJ Bird.	Assist passage of Team C. Secure lanes.	Reduce
Attack on OBJ Cat	Orient TRP 33.	On order, lift fires and occupy CP 16. Identify lanes.	On order, seize OBJ Cat.	Secure/improve lanes.	
Attack on OBJ Dog	Lift fires. Consolidate on SBF 25.	On order, seize OBJ Dog.	Occupy SBF 45. Orient TRP 33. On order, lift fires.	Assist passage of Team B. Secure/improve lanes.	Action on OBJ
Consolidate	SBF 25. Orient north.	OBJ Dog. Orient northeast.	OBJ Cat. Orient east.	Support hasty defense operations.	

Figure 1-4. Sample Execution Matrix

Effective C^2

1-50. Effective C^2 is paramount to mission success. C^2 is integrated into the plan through the use of maneuver and fire control measures and the positioning of key leaders to see the battle space. Maneuver control measures enable the commander to graphically convey his intent, scheme of maneuver, and subunit instructions. Relating subunit actions to the terrain is critical to successful execution.

1-51. In *Figure 1-3*, page 1-15, the commander uses PLs to initiate the breaching operation and to control actions on the objective. At PL Austin, the support force maneuvers to occupy SBFs 15 and 25. The assault force occupies checkpoint (CP) 16 to prevent premature advancement to the obstacle, reduction-site congestion, and the engagement of forces before lanes are created and marked. The commander uses target reference points (TRPs) on obvious terrain features to orient, focus, and shift suppressive direct and indirect fires.

1-52. Key leaders must be able to see the battle space to make informed decisions. Nowhere is this more true than during the breaching operation. The commander must position himself where he can best control the battle. Since effective suppression is the most critical event during breaching operations, the commander may position himself with the support force. This enables him to personally influence fire control and facilitate the necessary cross talk between breach and assault forces. The S3 may initially move with the breach force to track the progress of obstacle reduction and anticipate the commitment of the assault force. A commander who feels that his personal influence is required with the breach or assault force must make a conscious effort to track the entire battle and must not focus completely on the reduction of the obstacle or the assault itself.

Well-Rehearsed Forces

1-53. The most effective synchronization tool available to the commander is the rehearsal. The inherent complexity of the breaching operation makes rehearsals at every level essential to success. The commander must give his subordinates time to plan how they will execute their assigned missions and time to rehearse the plan with their unit. The TF facilitates rehearsals by preparing rehearsal sites and managing their use. The TF commander must be sensitive to the impact TF-level rehearsals have on subordinate planning and preparation, and he must carefully choose the time and place of the rehearsal. Company teams rehearse their support, breach, and assault roles. TF rehearsals focus on synchronizing the maneuver of support, breach, and assault forces to achieve the breaching fundamentals. Rehearsals also highlight key events that must be coordinated during breach execution.

Typical Breaching Operations

1-54. *Table 1-3*, page 1-18, shows the types of breaching operations versus the sizes of the enemy forces overwatching the obstacle.

Table 1-3. Types of Breaching Operations Versus Enemy Sizes

Enemy Overwatching the Obstacle	Maneuver Units Conducting Breaching Operations		
	Brigade	TF	Company
MIBN	X		
MIC	O	X	
MIP		O	X

NOTES:
X = Normal (combat-power ratio 3:1)
O = Desirable (combat-power ratio >3:1)

BREACHING OPERATIONS IN SUPPORT OF DELIBERATE AND HASTY ATTACKS

1-55. Attacks take place along a continuum based on the knowledge of the enemy's capabilities and disposition and the intentions and details of the friendly force's planning and preparation. The terms *deliberate attack* and *hasty attack* refer to the opposite ends of that continuum and describe characteristics of the attack:

- **Deliberate attack**. Friendly forces have more knowledge of the enemy situation and have more time to plan and prepare for the enemy. With this knowledge, friendly forces conducting the deliberate attack can develop detailed plans, task-organize to accomplish the mission, and execute rehearsals down to the lowest levels.
- **Hasty attack**. The enemy situation is vague, or it changes rapidly. Friendly forces have little or no time to plan or prepare for this type of operation. Because the primary purpose of the hasty attack is to maintain momentum or take advantage of the enemy situation, it is normally conducted only with the resources that are immediately available.

1-56. Breaching operations may be required to support an attack anywhere along the continuum. During a deliberate attack, friendly forces, such as engineers and smoke platoons, are task-organized before the attack to best support the concept of the operation. In a hasty attack, there may be insufficient time to task-organize assets, such as engineers and reduction assets, in the best manner. Additionally, the unit may not have time to plan the breaching operation in detail but may have to rely on the unit's SOP for execution. Regardless of where the attack falls along the continuum, the breaching tenets apply when conducting the breaching operation in support of the attack.

PLANNING SEQUENCE

1-57. The scheme of maneuver, engineer operations, fires, air defense, and actions at obstacles are all based on the same SITEMP. Planners develop the breaching plan using the following sequence:

- Reverse planning begins with actions on the objective.

- Actions on the objective drive the size and composition of the assault force.
- The size of the assault force determines the number and location of lanes to be created.
- Lane requirements and the type of obstacle drive the amount and type of mobility assets task-organized to the breach force.
- The ability of the enemy to interfere with the reduction of the obstacle determines the size and composition of the security element in the breach force.
- The ability of the enemy to mass fires on the point of breach determines the amount of suppression required and the size and composition of the support force.

Chapter 2

Breaching Operations

For a breaching operation, a unit develops a scheme of maneuver specifically designed to cross an obstacle and continue a mission. Maneuver company teams, TFs, and brigades can conduct breaching operations. Normally, a TF executes a breaching operation and the company teams are assigned as support, breach, and assault forces. Units will conduct a breaching operation when—

- The force-allocation ratio indicates that a confirmed enemy situation is beyond the subordinate unit's capabilities to successfully conduct a breaching operation. (When companies cannot successfully conduct a breaching operation, the battalion will.)
- A subordinate unit has failed in its attempt to breach an enemy's tactical obstacles.

NOTES:
1. See *Appendix D* for a breaching scenario and special planning considerations for breaching operations in restricted terrain.
2. See *Appendix E* for information on route clearance operations.

PLANNING

2-1. Units develop plans to conduct a breaching operation using the following military decision-making process (MDMP):

Step 1. Receipt of mission.
Step 2. Mission analysis.
Step 3. COA development.
Step 4. COA analysis (war game).
Step 5. COA comparison.
Step 6. COA approval.
Step 7. Orders production.

The following paragraphs highlight key portions of the MDMP as it applies to the planning of a breaching operation. For a complete discussion of the MDMP, see *FM 101-5*.

RECEIPT OF MISSION

2-2. Before receiving a higher HQ order, a unit should receive a series of warning orders (WOs) and available intelligence products. If the products are not issued with the WOs, the higher HQ liaison officer (LO) should locate and deliver the products as soon as they become available. Intelligence products are essential for the unit to begin developing its own intelligence products and to facilitate parallel planning.

2-3. Upon receiving the higher HQ SITEMP, the unit develops standard terrain products using TerraBase II or a similar computer-software program. (For examples of TerraBase products, see *FM 5-71-3*.) This set of terrain products is called a terrain-visualization mission folder. (For more information on this subject, see *FM 5-105*.) Terrain products that support the planning of a breaching operation should be included in the unit's SOP and may include—

- Visible area plots from each of the templated enemy locations to the expected EA.
- Visible area plots from potential SBF positions and points of breach to the templated enemy location.
- Oblique view of the AO.
- Perspective view of intervisibility lines.
- Visible area plots from named areas of interest (NAIs), when NAIs are developed.

2-4. Visible area plots from each of the templated enemy locations help determine if the enemy would select these locations. For example, the enemy would not select a location directly behind a small hill because the hill would prevent the enemy from firing into a kill sack. Additionally, by combining the visible area plots on an overlay, the enemy's ability to mass fires is more easily understood. It also helps select SBF locations, obscuration locations, and sectors of fires and helps determine the breach organization for future operations.

2-5. Visible area plots from potential SBF positions and points of breach assist future planning by allowing the commander to visualize the support and breach forces' ability to suppress the enemy. Multiple plots along the length of an SBF position may be necessary to fully understand the suitability of an SBF position.

2-6. An oblique view of the AO helps the commander better visualize the terrain. A perspective view of intervisibility lines in the AO allows him to visualize how the unit can best use the terrain during a breaching operation. It also provides an idea of how the enemy may use the intervisibility lines when positioning its assets and special munitions.

2-7. Visible area plots from developed NAIs can be used to provide positioning guidance for observation posts (OPs). These products are issued to recon assets who are tasked to provide information on various NAIs.

2-8. When receiving a mission, pay particular attention to the actions the higher HQ takes in support of the breaching operation or its decisions that influence the breaching operation. Responsibilities of the higher HQ include—

- Resourcing the unit conducting the breaching operation with additional or special assets not normally task-organized to subordinate units.
- Fixing enemy forces to prevent them from repositioning and interfering in the breaching operation.

- Using scatterable mines (SCATMINEs), deep fires, attack helicopters, and close air support (CAS) to isolate the farside objective and prevent a successful enemy counterattack.
- Planning a forward passage of lines by follow-on forces through the unit conducting the breaching operation. This must include a plan to upgrade/maintain lanes through the obstacle and to create additional lanes, if necessary, for the passage of follow-on forces.
- Providing criteria to transition to its own breaching operation.

MISSION ANALYSIS

2-9. During mission analysis, the commander conducts the IPB. A product of the IPB process is the SITEMP. The SITEMP is a graphic depiction of expected threat disposition if it adopts a particular COA. The staff must combine higher HQ intelligence with its knowledge of the enemy to produce a SITEMP with the necessary detail to successfully plan a breaching operation. A SITEMP should include—

- Likely enemy EAs.
- The location and orientation of enemy forces.
- Counterattack objectives and the location of enemy reserve elements.
- The location and range of all direct- and indirect-fire systems (tanks, AT weapons, artillery).
- Enemy obstacle systems, including tactical and protective obstacles and SCATMINEs. The template must depict the enemy's countermobility capability.
- The enemy's use of nuclear, biological, chemical (NBC) weapons, including the ranges of delivery systems.
- The location of enemy target-acquisition assets.
- Likely air avenues of approach (AAs), including fixed and rotary-wing aircraft.
- Positioning of enemy air-defense artillery (ADA) assets.

2-10. Selecting the initial commander's critical-information requirements (CCIR) is a part of mission analysis, and it is crucial to begin planning recon operations that will result in useful intelligence, including OBSTINTEL. Typical CCIR for a breaching operation may include the—

- Location, composition, and orientation of the obstacle and available bypasses in the vicinity of the reduction area and the point of penetration (PIR).
- Location and composition of enemy forces that are capable of employing direct and indirect fires on the point of breach (PIR).
- Maintenance status and location of all reduction assets (friendly forces information requirements [FFIR]).
- Status of commitment criteria for the breach force (PIR and FFIR).

NOTE: The initial CCIR is developed during COA analysis.

COURSE-OF-ACTION DEVELOPMENT

2-11. When developing a COA, the commander must remember that the breaching operation is only a portion of the higher HQ operation. The unit must ensure that the end state of the COA supports the higher HQ concept of the operation. As stated in *FM 101-5*, one of the criteria that each developed COA must meet is completeness. In a breaching operation, units use detailed reverse planning, develop a plan to achieve the breaching fundamentals, develop clear subunit instructions to develop a complete COA, and begin to synchronize the operation.

2-12. The detailed reverse-planning process, the SITEMP, and the terrain products produced during mission analysis are used to determine the size and composition of support, breach, and assault forces. The reverse-planning process begins with actions on the objective and includes templated enemy obstacles.

2-13. The assault force's task and purpose drive its composition. The assault force's mission is to seize terrain on the farside of the obstacle to prevent the enemy from placing direct fires on created lanes. Consider the following when determining the assault force's size and composition:

- Higher HQ requirements in terms of seizing or securing objectives.
- The terrain.
- The expected disposition, composition, and strength of the enemy when the assault force is committed.

2-14. The assault force's size determines the number and location of lanes to be created. The breach force creates and marks lanes through or around obstacles to support the rapid passage of the assault force. An assaulting company team requires a minimum of one lane, and an assaulting TF requires at least two lanes.

2-15. Lane requirements and the type of obstacle drive the amount and type of reduction assets task-organized to the breach force. As stated in *Chapter 1*, the commander should plan for the loss of about 50 percent of his reduction assets during a breaching operation. To ensure that the breach force is adequately resourced, the initial allocation of assets is continually reviewed as additional intelligence information (including OBSTINTEL) is received.

2-16. The enemy's ability to interfere with obstacle reduction determines the size and composition of the security element in the breach force. The security element's size and composition within the breach force can also be determined through the—

- Terrain analysis and the terrain's impact on the support force's ability to suppress the enemy.
- Amount of terrain the breach force will occupy at the obstacle during reduction and the extent that the breach force will mask the fires of the support force.
- Analysis of the enemy's expected reaction to the reduction effort.

2-17. Detailed terrain analysis and the enemy's expected size, disposition, and composition determine the enemy's ability to mass fires at the point of breach. That ability, in turn, determines the amount of suppression that is required and the support force's size and composition. Expected losses that the support force will incur from the line of departure (LD) to the SBF position due to maintenance and enemy air, chemical, and artillery attacks should be considered. Additionally, analyzing the enemy's equipment type, range, and lethality can indicate the expected losses that the support force will incur while in an SBF position.

2-18. When developing COAs for a breaching operation, consider the following:

- **Suppress.**
  - Identify the support force's axis of advance to SBF positions. The support force's mobility requirements must be identified based on the SITEMP and its axis of advance to SBF positions.
  - Establish SBF positions for each element in the support force.
  - Establish TRPs to control the support and breach forces' direct fires.
  - Determine the location and size of CFZs to protect the support force.
  - Plan essential indirect-fire targets to assist in suppressing the enemy.
  - Develop a plan for lifting or shifting direct and indirect fires.
  - Deploy the support force into a position to fire simultaneously on the enemy from several directions, if possible.
  - Attack by fire before or during obstacle reduction to suppress an overwatching enemy. If circumstances allow, forces may bypass the obstacle and assault the overwatching enemy before reduction. If the assault successfully eliminates the enemy, it also eliminates direct fires on the breach force. If the assault does not succeed in destroying the enemy, it fixes them and reduces the enemy's ability to place fires on the breach force. Direct- and indirect-fire suppression from the support force is necessary to support this assault.
  - Use fires to suppress the enemy's indirect fires. The enemy's indirect fires are extremely dangerous during a breaching operation since the breach force may be exposed for a lengthy period. Friendly artillery and air assets support the breaching operation by attacking enemy overwatch positions, mobile reserves, and artillery positions that are capable of firing on the point of breach. Friendly forces jam enemy fire control nets. The fire-support officer (FSO) designates points of breach as CFZs for a target-acquisition battery so that enemy artillery is treated as a priority target for counterbattery support.
- **Obscure.**
  - Determine the location of smoke target(s).
  - Identify terrain features that mask the force's point of breach, and limit the number of enemy forces that can effectively observe and fire on the breach force, thus aiding in obscuration. Other natural, limited-visibility conditions (darkness, fog) should be exploited if possible. Smoke is generally the primary obscurant; however, it is

a double-edged sword because it may attract enemy attention and degrade friendly target acquisition and C^2.

- Determine the quantity, location, and type of obscurant required to screen breach and assault forces and obscure enemy fires. This is particularly important for breaching complex obstacles where the breach force is exposed for an extended time. Smoke is important in deceiving the enemy as to the breaching location and interfering with enemy fires. If smoke platoons are available, synchronize them with artillery-delivered and organic smoke.
- Determine smoke requirements by using *FM 3-50*. Blanket, haze, or curtain-screening smoke degrades enemy observation of assault and breach forces. A smoke platoon is capable of providing support for a 0.5- to 1.4-kilometer front. Friendly forces use a combination of carefully placed smoke pots and indirect fires to augment smoke support. Howitzers and mortars supporting friendly forces can build and maintain smoke, but smoke effectiveness depends on the number and rate of smoke rounds delivered. Using indirect fire to provide smoke competes with the friendly force's high-explosive (HE) and dual-purpose improved conventional munition (DPICM) missions against an enemy. A commander must decide the priority for the use of indirect fire, and his decision must be clear to subordinates. Vehicles equipped with smoke-grenade launchers can provide local smoke for reduction operations and for troop passage during an assault. However, these systems are self-defense mechanisms and friendly forces must not rely on them to supply all obscuration for a breaching operation.

- **Secure**.
  - Plan direct-fire control measures as necessary.
  - Use terrain and the positioning of nearside and farside security SBF near the point of breach to facilitate the survivability of the breach force.
  - Plan a CFZ and ADA coverage around the reduction area for survivability and security of the breach force.
  - Plan local security at the breach site to suppress overwatching enemy forces that are close to the obstacle.
- **Reduce**.
  - Identify primary and alternate reduction techniques.
  - Determine the tentative location of lanes and the separation between lanes.
  - Determine the breach force's task organization.
  - Position the breach force, and identify positions and routes that it uses as it maneuvers toward the obstacle to minimize its impact on hindering the support force's suppressive fires.
- **Assault**.
  - Plan to pass the assault force through lanes created in the obstacle.
  - Develop objectives for the assault force.

- Develop SBF positions to be occupied by the assault force to facilitate passage of follow-on forces.
- Plan direct- and indirect-fire control measures that support the seizure of the farside objective.
- Identify all mobility requirements for the COA, including whether the assault force requires engineer support for reducing protective obstacles and fortifications within the defensive position.

2-19. Each element within the breach organization must completely understand which tasks it is responsible for in breaching operations. Additionally, each COA should identify which force is the main effort throughout the breaching operation. For example, the COA should define when the main effort shifts from the support force to the assault force. The following are the different types of forces and their tasks and responsibilities:

- **Support force**.
  - Suppress enemy elements capable of placing direct fires on the point of breach to protect the breach force.
  - Suppress enemy elements capable of placing direct fires on the assault force. This may require the adjustment of SBF positions.
  - Call for and adjust indirect fires, including obscuration. A combat observation lasing team (COLT) or Striker will be required for Copperhead munitions.
  - Fix enemy forces that are capable of repositioning.
- **Breach force**.
  - Reduce lanes in the obstacle.
  - Provide local security (farside and nearside).
  - Provide additional suppression of enemy forces overwatching the obstacle.
  - Mark and report the location of created lanes.
  - Assist the passage of the assault force through created lanes.
- **Assault force**.
  - Seize the farside objective.
  - Reduce protective obstacles.
  - Prevent enemy direct fire from interfering with follow-on forces as they pass through lanes.
  - Provide clear routes to the BHL for follow-on forces.
  - Conduct battle handover with follow-on forces.

2-20. The assault force's mission is unique. It must reduce the enemy's protective obstacles, assault the enemy's position, and secure terrain that is essential to the passage of follow-on forces. The assault-force commander must thoroughly understand the higher HQ scheme of maneuver. The following information is particularly important:

- Objectives the assault force must seize.
- The end state, in terms of SBF positions, which the assault force must occupy so that follow-on forces can continue the attack.

- The number of lanes the breach force plans to create to facilitate the passage of the assault force.
- The higher HQ plan to provide suppressive fires and/or obscuring smoke.
- Planned points of breach and the point of penetration.
- Lane requirements (number and lane-marking type) to support the passage of follow-on forces.
- The sequence of follow-on units to be passed through lanes and the criteria to begin passing follow-on units through lanes that the assault force created.

2-21. Typically, the enemy emplaces protective obstacles 50 to 500 meters in front of its positions and between its forward platoons and companies. These obstacles are designed to prevent a mounted assault and to fix or break up a dismounted assault. Close-in obstacles may be a combination of wire, AP and AT mines, fortifications, and entrenchments. All obstacles are covered by direct fire. Mines may be surface-laid or buried and well-camouflaged. Mines can be deployed with AHDs that are command-detonated or trip-wire-activated.

2-22. The assault force commander considers the following breaching fundamentals when developing his plan:

- **Suppress**. If a company team is the assault force as part of a TF breaching operation, other company teams within the TF will normally have the responsibility to suppress the enemy from designated SBF positions on the objective area. These teams may be the same support force that protected the breach force; however, terrain factors may require repositioning to provide effective support for the assault force.
- **Obscure**. Mortars and vehicle-mounted obscurants are primary means of obscuration for the assault force. It may also employ hand-held smoke grenades and smoke pots to provide additional obscuration. If available, smoke may be employed away from the point of breach to confuse enemy forces as to the disposition of the assault force.
- **Secure**. Local security must be established before conducting reduction operations on protective obstacles.
- **Reduce**.
  - The assault force's reduction efforts center on creating dismounted/mounted lanes in protective obstacles, reducing fortifications, and widening lanes to support follow-on forces. The commander should task-organize the assault force with sufficient reduction assets to create a minimum of one lane per assaulting platoon. There may be insufficient time to reorganize engineers and reduction assets to the assault force after crossing the LD. Therefore, engineers and reduction assets supporting the assault force are task-organized to the maneuver force they will support.
  - During the planning process, the assault force commander must decide whether he will assault mounted or dismounted. This decision impacts the reduction methods that will be used. A

mounted assault allows the use of tank plows and MICLICs. During a dismounted assault, friendly forces rely on manual obstacle reduction, which is very time-consuming. The composition of follow-on forces must also be considered. This may require widening initial lanes that the assault force created, especially when a dismounted assault is followed by mechanized follow-on forces.

  - The assault force normally decentralizes C^2 for the plan's execution. For this reason, an assault force commander must allocate engineers, reduction assets, and demolitions to squads and platoons to provide reduction capability throughout his force.

- **Assault**. Below-ground operations require clearing the enemy's trench lines and bunker complexes. Dismounted infantry and engineers, in concert with tanks and Bradley fighting vehicles (BFVs), conduct these operations. These actions must be planned in detail and rehearsed to ensure proper execution. See *FMs 71-1* and *7-10* for more information.

COURSE-OF-ACTION ANALYSIS (WAR GAME)

2-23. When planning a breaching operation, war gaming is critical to refine the COA, including—

- Task-organizing support, breach, and assault forces.
- Developing a DST, identifying triggers to support obscuration, committing breach and assault forces, and shifting fires.
- Identifying and/or confirming the locations of decision points, NAIs, targeted areas of interest (TAIs), and other information that is needed to support decision points.
- Refining CCIR and incorporating them into the ISR plan/graphics.
- Refining the fire-support plan/graphics, including the obscuration plan.
- Confirming the obstacle template according to the TF R&S plan.

2-24. The commitment of the breach force is a critical decision point that must be included in the DST. The commander must thoroughly analyze what criteria must be met before the breach force begins obstacle reduction. Commitment criteria elements may include the following:

- Destruction of certain vehicles or a certain number of vehicles.
- Effective suppression of the enemy by the support force.
- Effective obscuration of the enemy.
- Remaining strength of the support force.
- Remaining reduction assets available to the breach force.
- Activation of CFZ.
- Air-defense assets in position.

NOTE: A plan must exist on how information will be gathered to verify the commitment criteria.

2-25. Once war gaming is complete, a COA has been refined, and a final task organization has been determined, a C^2 system can be planned to support the operation. To conduct a successful breaching operation, an effective C^2 system must be established. Specifically, each force in the breaching organization (support, breach, and assault) must understand who is in charge, especially within support and breach forces.

2-26. The support force may be comprised of more than one element. For example, in a TF breaching operation, the support force may be comprised of two company teams. Due to the importance of the support force's role and its size, it may be best to have the TF commander/S3 be the support force commander. He is responsible for coordinating the actions of both company teams and determining when the conditions have been met for the breach force's commitment. This allows each company commander to concentrate on directing his unit while under the overall control of an experienced leader.

2-27. The breach force, with its two subordinate elements (security and reduction), must have a well-understood C^2 system in place. There are many ways to establish C^2 of the breach force, and the method used may depend on the—

- Size and composition of security and reduction forces.
- Number of lanes that must be created.
- Experience level of the personnel within the breach force.

2-28. The following are possible C^2 structures within the breach force for a TF breaching operation:

- A company-team commander acts as the security element commander and the breach force commander, with an engineer as the reduction element commander.
- An engineer-company commander acts as the reduction element commander and the breach force commander, with a maneuver platoon leader as the security element commander.

2-29. Another C^2 issue that must be resolved during COA analysis is traffic control through lanes created in an obstacle system. *Figure 2-1* shows an example of the graphic control measures that may be established to facilitate the efficient movement of forces through an obstacle system. In this example, traffic-control post (TCP) 8 is located at the far-recognition marker. These control measures can be initially established by the higher HQ but need to be refined as the unit conducting the breaching operation develops its scheme of maneuver. *Table 2-1* lists the responsibilities for creating lanes and controlling traffic through an obstacle system when a TF breaching operation is being conducted.

PREPARATION

2-30. Intelligence collection continues throughout preparations for a breaching operation. As intelligence reports are received, the SITEMP and the ISR plan are updated and revised. Intelligence reports are also used to refine the task organization of support, breach, and assault forces and the scheme of maneuver, including proposed points of breach and target grids for smoke and suppressive fires. Updated intelligence information is also used during combined-arms rehearsals.

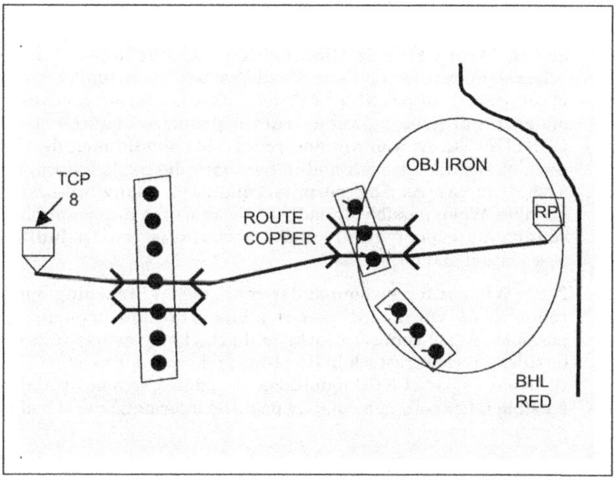

Figure 2-1. Graphics to Support Passage of Follow-On Forces

Table 2-1. C^2 Responsibilities for Lanes and Traffic Control

Breaching Operation	Responsibilities of TF Forces	Responsibilities of the Brigade
Initiating breaching operations through the passage of the assault force through tactical obstacles	The breach force creates lanes for the assault force (reduces, proofs, and emplaces the initial lane-marking pattern and assists in the passage of the assault force through the marked lanes). The TF controls all traffic passing though tactical obstacles.	
Completing the passage of the assault force through the seizure of the farside objective	The assault force creates and marks lanes through protective obstacles up to the BHL.	Assumes control of lane(s) in tactical obstacles Creates more lanes as necessary Upgrades lane marking to the intermediate lane-marking pattern Ensures that lanes remain open Assumes traffic control through tactical obstacles
Passing follow-on forces through tactical and protective obstacles to the BHL	The assault force assists in the passage of follow-on forces (from tactical obstacles to the BHL).	
Completing the forward passage of lines		Assumes control of all lanes and traffic up to the BHL

2-31. A unit meticulously plans, manages, and controls breaching rehearsals. (For more information on breaching rehearsals, see *Chapter 4* of this manual or *FM 101-5*.) Time is allocated to each unit to perform combined-arms rehearsals. A rehearsal site should reflect the actual obstacle system in as much detail as possible. Friendly forces choose terrain similar to the operational area and construct a practice obstacle system based on OBSTINTEL. At a minimum, rehearsals should include a leader and key personnel walk-through and individual rehearsals by support, breach, and assault forces. As time permits, conduct as many full-scale rehearsals as possible. When possible, friendly forces rehearse the operation under the same conditions expected during the actual mission, including battle-space obscuration and darkness.

2-32. When a force commander rehearses a breaching operation, he also rehearses several contingency plans. The contingencies should include possible enemy counterattacks and attacks by enemy indirect-fire systems (artillery, rockets, attack helicopters, and other air assets). This also includes the enemy's use of NBC munitions. If updates become available after the last possible rehearsal, immediately pass the information to the affected elements.

EXECUTION

2-33. Each element of the breaching organization must accomplish its assigned tasks for a breaching operation to be successful. Additionally, the entire force must be flexible when responding to changes in the plan. Examples of changes are—

- Allocating additional assets to the support force due to attrition.
- Changing the location of the point of penetration.
- Modifying the scheme of reduction and changing the order of the units passing through the created lane.

2-34. All units must continually update the commander during mission execution. A critical phase of a breaching operation is the movement of the support force from the LD to its SBF positions. The support force must be prepared to maneuver to its SBF positions. Additionally, the support force must plan for the enemy's use of SCATMINEs along its axis of advance, which may require it to conduct a breaching operation.

2-35. After the support force has occupied its SBF positions and the commitment criteria of the breach force has been met (achieved necessary suppression and obscuration), the breaching commander should order the breach force to move forward to a specific point (6- or 8-digit grid) on the battle space to begin reduction. As the breach-force commander maneuvers his force to the point of breach, he must always remain cognizant of his relationship to the support force to ensure that he does not unnecessarily mask their fires.

2-36. Actions by the assault force at an assault position may include—

- Verifying current friendly and enemy situations using tactical reports from company teams or support forces.
- Issuing fragmentary orders (FRAGOs) and disseminating information to the lowest level.

- Confirming TRPs and direct-fire responsibilities.
- Positioning artillery observers.
- Conducting final prepare-to-fire checks.
- Reorganizing to compensate for combat losses.

2-37. The breach force commander should lead with his security element to allow it to establish nearside security before the reduction element arrives at the obstacle. As the security element occupies its position, the reduction-element commander should conduct a recon to acquire obstacle information and quickly confirm/refine his scheme of reduction. As the reduction-element commander brings his element forward, he should report the exact location where the obstacle will be reduced and the estimated time to reduce/proof and mark the obstacle. At the same time, the breach-force commander must assess the effectiveness of the suppression and determine if and how he will augment the support force's fires.

2-38. As the reduction continues, the support- and breach-force commanders must update their commander with the current status. Specifically, the support-force commander must update on his ammunition status and the strength of his unit. The breach-force commander must update on the progress of the reduction effort and the effectiveness of obscuration and suppression. The breach-force commander should verify who the assault force will be and where the assault force will come from so that he can assist in their passage.

2-39. As the reduction effort nears completion, the breach-force commander should report the grid coordinates of the far-recognition marker so that the assault force can begin movement to that location. When reduction is complete, the breach-force commander must—

- Report lane completion.
- Provide grid coordinates of the final-approach marker.
- Confirm/report the lane-marking pattern and material.

2-40. The breach-force commander establishes farside local security and assists in the assault force's passage. Once the lanes have been reported/signaled as being open for traffic, the assault force passes through the lanes to destroy forces that can bring direct fires on created lanes. To accomplish this mission, the assault force may have to reduce enemy protective obstacles.

2-41. The obstacle system continues to be a choke point and danger area even after the assault force has passed through created lanes. Additional lanes are constructed to speed the follow-on forces' passage, and lanes are monitored to ensure that they remain clear. Reduction assets need to be maintained near the points of breach in case a scatterable minefield is emplaced by the obstacle. The lane-marking pattern is upgraded to intermediate. There are many ways this can be accomplished. The higher HQ—

- Takes command of the point of breach and has some or all of the reduction element revert to its control and continue creating/expanding lanes in the obstacle for follow-on forces.

- Takes command of the point of breach and has additional reduction assets follow closely behind the assault force and create additional lanes to allow the reduction element to remain under unit control.
- Tasks the breaching unit to maintain lanes in tactical obstacles.

2-42. The lanes are widened to allow two-lane traffic through obstacles, and they are marked with the full lane-marking pattern. Deliberate marking and fencing systems are installed, and military police (MP) establish necessary traffic control. Eventually, follow-on engineer forces clear obstacles and eliminate the choke point.

Chapter 3

Lane and Bypass Marking

This chapter implements STANAG 2036.

This chapter provides commanders with an Army-wide standard system for lane and bypass marking. It centers around a systematic, phased upgrade of lane marking. Each upgrade conforms to tactical requirements for that phase of the attack, from initial obstacle reduction to the passage of larger follow-on forces and the return traffic necessary to sustain the force. Lane or bypass marking is a critical component of obstacle reduction. Effective lane marking allows commanders to project forces through an obstacle quickly, with combat power and C^2 intact. It gives an assault force and follow-on forces confidence in the safety of the lane and helps prevent unnecessary casualties.

LANE-MARKING SYSTEM

3-1. There are two critical components to any lane-marking system—the lane-marking pattern and the marking device.

- **Lane-marking pattern**. The location of markers indicating the entrance, the lane, and the exit.
- **Marking device**. The type of hardware emplaced to mark the entrance, the lane, and the exit.

3-2. The lane-marking system outlined in this chapter centers on standardizing the marking pattern across the Army rather than standardizing the marking device. Standardizing the marking pattern is critical to offensive operations. A common lane-marking pattern—

- Enables cross attachments and adjacent units to recognize lanes easily with minimal knowledge of a particular unit's tactical SOP.
- Gives all forces a standardized set of visual cues needed to pass through a lane safely and maintain their momentum.
- Facilitates quick conversion to the lane-marking requirements of STANAG 2036.

3-3. Until the Army adopts a standard lane-marking device, commanders decide what hardware to use. This gives units greater flexibility, allowing them to adopt a marking device that is tailor-made for their type of unit and mission (armored or light force, mounted or dismounted attack, limited visibility, thermal capability). Regardless of the device used, it must support the standard lane-marking pattern outlined in the following paragraphs. Therefore, commanders should consider the marking guidelines and examples in this chapter before developing or adopting their own system. Commanders should determine and standardize the marking device used based on the availability of resources and METT-TC.

LANE-MARKING TERMS

3-4. The following terms provide a common basis for discussing lane marking:

- Entrance markers.
- Handrail markers.
- Exit markers.
- Entrance-funnel markers.
- Final-approach markers.
- Far-recognition markers.
- TCPs and guides.

ENTRANCE MARKERS

3-5. Entrance markers indicate the start of a lane through an obstacle. They signify the friendly-side boundary of the obstacle and the point at which movement is restricted by the lane width and path. Entrance markers also indicate the lane width. Therefore, it is critical for a maneuvering force to distinguish the entrance point clearly, since it signifies the point at which passing vehicles can no longer adjust their movement in reaction to the situation (direct and indirect fires) without jeopardizing the force. Entrance markers must be visually different from handrail markers to help a force distinguish this critical point in the lane. The distance between entrance markers must be the same as the width of the created lane. Entrance markers placed a minimum of 4.5 meters apart indicate a lane capable of supporting mounted movement, while markers placed a minimum of 1 meter apart indicate a dismounted lane. When obstacle boundaries are vague or unknown, such as a buried minefield, the breach force uses its best judgment and marks the entrance where obstacle reduction and lane proofing begin.

HANDRAIL MARKERS

3-6. Handrail markers define the lane path through an obstacle and indicate the limits of the lane width. As a minimum, mounted and dismounted lanes have a left handrail. Mounted and dismounted forces moving through a lane should keep the left handrail immediately to the left of their vehicle or person. The lane width is defined by the entrance markers. Therefore, when only the left handrail is marked, drivers use entrance and handrail markers to gauge the lane width and path. As the operation progresses, lane marking may be upgraded to include left and right handrails.

EXIT MARKERS

3-7. Exit markers indicate the far side limit of a lane through an obstacle. For a passing force, the exit marker signifies the point at which movement is no longer confined to the lane path. Like entrance markers, exit markers must be distinguishably different from handrail markers; however, the exit may be marked the same as the entrance. Exit markers are placed to the left and right of an exit point and spaced the width of the reduced lane. This visual reference is critical when only a left handrail is marked. The combination of entrance, left-handrail, and exit markers gives the driver and the tank commander visual cues (entrance and exit points, lane width and path) to pass safely along a lane.

ENTRANCE-FUNNEL MARKERS

3-8. Entrance-funnel markers augment entrance marking. The V formed by the funnel markers forces approaching platoons into a column formation and assists drivers and tank commanders in making last-minute adjustments before entering the lane.

FINAL-APPROACH MARKERS

3-9. Final-approach markers are highly visible, more robust markers that augment the visual signature of entrance-funnel markers. Units must be able to see the entrance funnel and/or the entrance-funnel markers from the final-approach marker. The final-approach marker—

- Provides the assault-force commander with a highly visible reference point toward which to maneuver his formation.
- Signals company-team commanders to begin changing to a column formation.

FAR-RECOGNITION MARKERS

3-10. Far-recognition markers are highly visible markers located between the final-approach marker and the friendly unit. They are primarily used when passing battalion-size forces through a lane where direct observation of the final-approach marker is denied due to distance, visibility, or terrain. Far-recognition markers should be different from the final-approach marker. Far-recognition markers indicate the point at which forces begin changing their formation to posture for the passage. A single far-recognition marker may serve up to two lanes when located 200 to 400 meters apart. Once lanes are upgraded to two-way traffic, far-recognition markers are required for each two-way lane. Far-recognition markers should be visually alterable so that they can be distinguished from the far-recognition marker of an adjacent lane. This assists the C^2 of large formations when passing on several adjacent lanes. When a far-recognition marker serves more than one lane, a guide or a TCP is collocated with the far-recognition marker nearest the obstacle.

TRAFFIC-CONTROL POSTS AND GUIDES

3-11. A TCP or a guide is a two-man team with communications means that assists the commander in controlling the movement of forces. When available, consider using MPs to establish and man these posts. Initially, the commander may use other personnel as guides to man critical far-recognition markers until MPs establish full TCPs. TCPs and guides provide the commander with a means to control traffic flow to appropriate lanes. When there are multiple lanes branching off a single far-recognition marker, the TCP helps direct the formation to various lanes. The TCP can also assist in modifying the traffic flow when lanes have been closed for maintenance, for lane expansion, or by enemy SCATMINEs. The TCP or guide must give the commander the azimuth and distance to the final-approach marker, the device used for the approach marker, and the level of the lane-marking pattern. For light forces, guides may physically escort passing units from the far-recognition marker to the lane entrance. In short, the TCP allows the commander to make last-minute changes in the traffic flow, giving him increased flexibility to react to the enemy situation.

3-12. Another technique to facilitate the movement of the assault force to the points of breach is for the assault force to maneuver a recon element, such as a section, behind the breach force. This section observes the breach effort, identifies lanes, and leads the assault force to lanes after the breach is complete.

LEVELS OF LANE MARKING AND PATTERNS

3-13. There are three levels of lane marking—initial, intermediate, and full. Each lane-marking level provides an increase in lane signature and capability. Lane requirements change as a breaching operation matures from the passage of the assault force to the passage of larger follow-on forces. Initial lane-marking requirements are driven by the nature of the fight through the obstacle. Marking must be rapid, providing only the bare minimum signature needed to safely pass small units (company teams, platoons) that make up the assault force. This contrasts with lane requirements during later phases of an attack where larger units (battalion and above) are passed to subsequent objectives and mark improvements to be made. Two-way traffic becomes a priority for the simultaneous forward passage of combat units and return traffic (ambulances, empty supply vehicles) necessary to sustain the force. With the increase in traffic volume comes more diverse forces and levels of driver experience. Lane-marking limits must be clear to the most inexperienced driver or crewman; do not assume he has a knowledge of the unit's SOP. A fully developed lane must support two-way traffic and be completely marked.

3-14. Commanders must be aware of how the lane needs of a force change with the operation so that they can anticipate lane-marking and lane-capability requirements. Integrating the levels of lane marking into the overall breaching plan ensures that a unit's needs are satisfied. The forces necessary to mark and upgrade lanes must be allocated and tasked with that mission. The scheme of maneuver and the service-support plan are the basis for analyzing lane requirements. The following paragraphs describe lane-marking patterns in detail and provide guidelines on when a commander should upgrade lane marking and lane capability.

INITIAL LANE MARKING

3-15. The breach force emplaces the initial lane-marking pattern immediately after the obstacle is reduced and, if required, the lane is proofed. This is a signal to the assault-force commander that the lane is ready for traffic. Initial lane marking is kept to a minimum, focusing on the markings needed to pass immediate assault forces (company teams and smaller) through the lane to seize the point of penetration on the farside objective. Normally, the assault force can observe the lane as it is being marked and does not need the greater visual signature of mature lane marking. The initial lane-marking pattern contains—

- Entrance markers.
- Left-handrail markers.
- Exit markers.
- Entrance-funnel markers.
- A final-approach marker.

3-16. *Figure 3-1*, page 3-6, shows the initial lane-marking pattern. Entrance, left-handrail, and exit markers are the first markers that the breach force emplaces since these markers define the location and limits of the lane. These markers should be emplaced as follows:

- Entrance markers are placed to the left and right of the reduced lane's entrance point. They are spaced the width of the lane (at least 4.5 meters for mounted lanes and 1 meter for dismounted lanes).
- Left-handrail markers are placed at the left limit of the lane along the entire path. They are placed at 15-meter intervals for mounted forces and 5-meter intervals for dismounted forces. Commanders can modify the intervals based on the terrain, the visibility, the lane length, and the lane path. Additionally, commanders may choose to mark left and right handrails when the lane path is lengthy or unclear (through a complex obstacle).
- Exit markers are placed to the left and right of the lane at the far side of the obstacle (at least 4.5 meters for mounted lanes and 1 meter for dismounted lanes).

3-17. After entrance, left-handrail, and exit markers are placed, the breach force emplaces entrance-funnel markers and the final-approach marker. These markers enhance the visual signature of the lane entrance for the assault force. If the assault force is in a position to observe the obstacle reduction, the commander does not have to wait until entrance-funnel markers and the final-approach marker are set to commit the assault force. However, the breach force continues to install the necessary markers to improve the lane signature for larger or subsequent follow-on forces maneuvering toward the lane. Entrance-funnel markers and the final-approach marker are emplaced as follows:

- Entrance-funnel markers are placed at 15-meter intervals for mounted forces and 5-meter intervals for dismounted forces. They are placed diagonal to the lane entrance to form a 45-degree V as shown in *Figure 3-1*.
- The final-approach marker is centered on the lane and placed at least 200 meters from the lane entrance for mounted forces. For dismounted forces, the nature of the attack may preclude using a final-approach marker initially. Place a final-approach marker 30 meters from the entrance as soon as the mission allows. Commanders modify the recommended distance for the final-approach marker based on the terrain and the visibility.

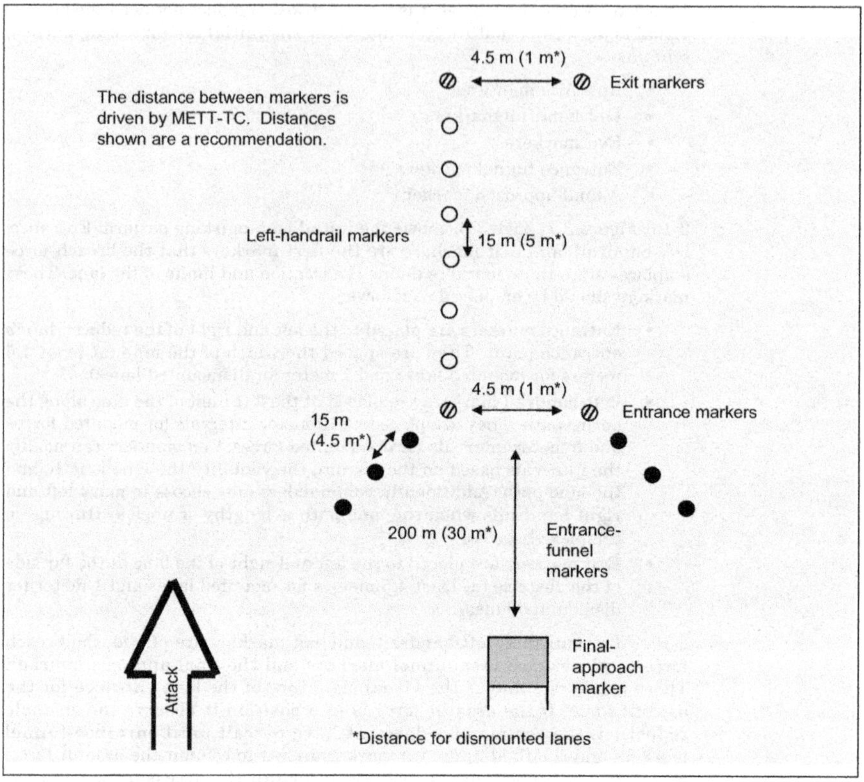

Figure 3-1. Initial Lane-Marking Pattern

INTERMEDIATE LANE MARKING

3-18. Upgrading initial lane marking to intermediate lane marking is triggered by one of two key events—

- Commitment of larger combat forces who are unable to directly observe the obstacle as it is reduced.
- Rearward passage of sustainment traffic (casualty evacuation and vehicle recovery).

3-19. Intermediate lane marking has two goals—

- Increasing the lane signature to assist in the passage of larger combat forces.
- Providing sufficient marking for two-way, single-lane traffic.

3-20. Intermediate lane marking builds on the initial lane-marking pattern by adding right-handrail markers, exit-funnel markers, far-recognition markers, and a farside final-approach marker. *Figure 3-2* shows the intermediate lane-marking pattern.

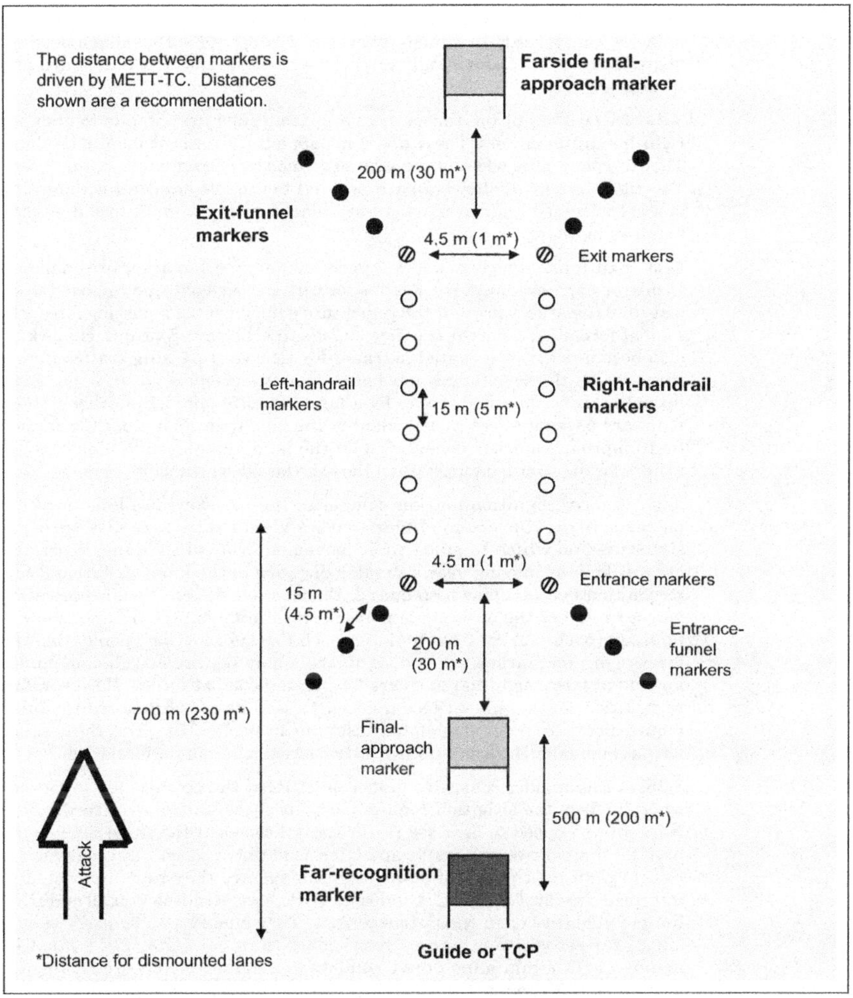

Figure 3-2. Intermediate Lane-Marking Pattern

3-21. The commander sets the priority of marker emplacement based on the situation. If the scheme of maneuver requires the immediate passage of larger combat forces, right-handrail markers and the far-recognition marker may be the priority. On the other hand, if it is necessary to ground-evacuate casualties or to recover vehicles, it may require emplacing right-handrail markers, exit-funnel markers, and the farside final-approach marker first. Intermediate lane marking gives the commander greater flexibility. Establishing a network of guides becomes increasingly important to direct the flow of forward and returning forces.

3-22. Regardless of the reason for an upgrade, the first step is to emplace right-handrail markers. These markers define the rightmost limit of the lane. They are placed the width of the lane as defined by entrance and exit markers. The right handrail follows a path parallel to the left handrail through the obstacle. Right-handrail markers are placed at the same interval as left-handrail markers.

3-23. Exit-funnel markers and a farside final-approach marker are emplaced to mirror entrance markers. For the forward passage of large combat forces, exit-funnel markers prevent the premature deployment of a passing force into combat formation before it is safely outside the obstacle. Exit-funnel markers also become entrance-funnel markers for rearward passing traffic, giving these forces the visual cues needed to align themselves on the lane. Exit-funnel markers are augmented by a farside final-approach marker to help rearward passing forces clearly identify the lane from their side. The farside final-approach marker is centered on the lane and placed 200 meters (30 meters for dismounted forces) from the exit (farside entrance).

3-24. A far-recognition marker completes the intermediate lane-marking pattern. It provides commanders with a visual signature or a series of signatures on which to guide their movement toward the lane. When the assault force is moving over a greater distance to the lanes, additional far-recognition markers may be required. For mounted forces, the far-recognition marker nearest the obstacle is placed a maximum of 500 meters from the final-approach marker (the final-approach marker must be visible from the far-recognition marker). Dismounted forces may require a system of guides instead of far-recognition markers for passing combat forces. However, far-recognition markers must be emplaced as soon as possible to reduce guide requirements for passing mounted sustainment traffic. This gives the assault-force commander the space needed to transition to a column formation.

3-25. A commander collocates guides or TCPs at the far-recognition marker when he feels the situation requires more positive control over traffic flow. Normally, the need to increase traffic control comes at the same time as the need for limited two-way traffic and intermediate lane marking. Commanders should plan for the use of dedicated guides once they have upgraded to intermediate-level marking. Guides or TCPs become mission critical during limited visibility or in restrictive terrain. They should also be used when a single far-recognition marker feeds more than one lane. TCPs must be manned with a minimum of two soldiers and must have secure frequency-modulated (FM) communications with the controlling HQ. It is essential that soldiers acting as guides or TCPs know the—

- Azimuth and distance to the final-approach marker or the 8-digit grid coordinate of the final-approach marker that is entered into the unit's Global Positioning System (GPS) receiver.
- Lane-marking pattern currently emplaced.
- Type of final-approach marker used.
- Traffic-control plan and march order.

Guides must be kept up-to-date on the status of lane marking, maintenance, and closures.

FULL (TWO-WAY) LANE MARKING

3-26. Lane marking is upgraded to a full lane-marking pattern when the commander desires uninterrupted two-way traffic flow. Expanding lanes to full lane marking is resource-intensive and is not part of a breaching operation. A full lane supports uninterrupted two-way traffic. The marking pattern is modified to give forces passing forward or rearward the same visual signature.

3-27. Upgrading to a full lane-marking pattern is normally assigned to follow-on engineer forces, since it is usually beyond the immediate capability of engineers with forward units. In special cases, the commander may task engineers with forward units to expand and upgrade a lane from an intermediate to a full lane-marking pattern. However, the trade-off is decreased support to forward units and delays associated with changing task organization. The full lane-marking pattern is also used when marking a lane through friendly obstacles along a main supply route (MSR) or a passage lane.

3-28. Upgrading intermediate lane marking to full lane marking begins by temporarily closing the lane, rerouting traffic, and expanding the lane width. The lane is always expanded to the left in relation to the direction of the attack. Engineers reduce and proof the obstacle beginning at the left handrail to give a total lane width of 10 meters (5 meters each way). The expansion width requirement is the same for armored and light forces, since both forces must be able to pass mounted sustainment and combat forces. When the lane is expanded by mechanical means, it is necessary to entirely remove the left-handrail markers during expansion. Whenever possible, these markers should be left alone to serve as a reference point during lane expansion. If removed altogether, left-handrail markers must be replaced since they form the left handrail for both forward and rearward passing forces.

3-29. When the lane width has been expanded to 10 meters, entrance, exit, right-handrail, funnel, and final-approach markers are replaced on the return lane. All markings are the same as described in previous paragraphs. *Figure 3-3*, page 3-10, shows a full lane-marking pattern.

3-30. The full lane-marking pattern has three entrance and three exit markers. Entrance and exit markers are placed the width of forward and return lanes and are visually different from other markers. Three entrance markers indicate a two-way traffic lane, and units should always use the rightmost lane. Entrance- and exit-funnel markers are placed slightly different from previous marking patterns. In the full lane-marking pattern, funnel markers extend out from the entrance and exit markers on the outside

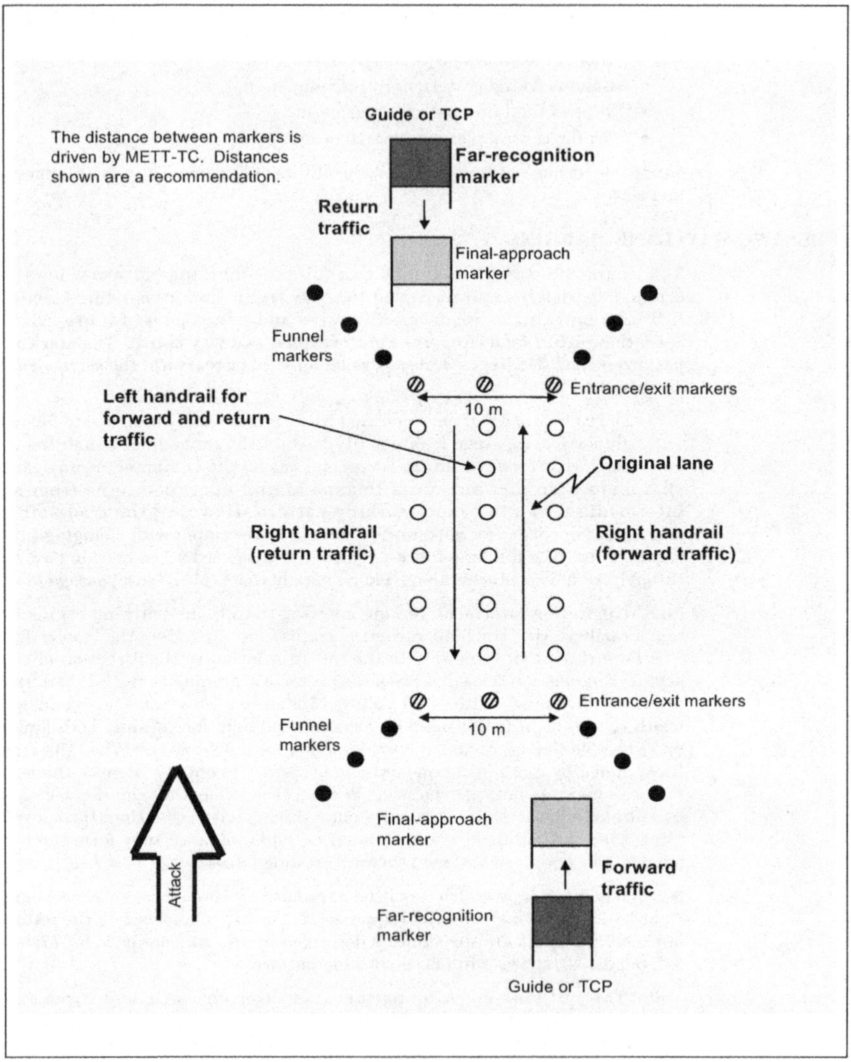

Figure 3-3. Full Lane-Marking Pattern

only. Final-approach markers are placed 200 meters from and centered on entrances of forward and return lanes. This helps forces identify entrance points from either direction. Far-recognition markers are placed a maximum of 500 meters from forward and return final-approach markers. Again, distances may be modified based on the terrain, the visibility, and the mission.

COMMANDER'S GUIDANCE FOR LANE MARKING

3-31. *Table 3-1* provides the commander with a summary of lane-marking levels, guidelines on unit responsibilities, and events that trigger a lane-marking upgrade. The chart focuses on TF-level breaching operations. In the table, *who* refers to the unit responsible for lane marking, *when* describes events that trigger the need to upgrade, and *markers* highlight a quick reference for lane-marking requirements.

Table 3-1. Lane-Marking Levels, Unit Responsibilities, and Trigger Events

	Initial	Intermediate	Full (Two Way)
Who	TF breach force	TF breach force	Brigade
When	Obstacle is reduced	Passing battalion- or company-size forces	Passing brigade- or battalion-size forces
	Passing platoon- or company-size assault forces	Passing force which cannot see the lane	Situation requires uninterrupted sustainment traffic
		Passing TF combat trains	
Markers	Entrance	Add right handrail	Expand lane width to 10 meters
	Exit	Add exit funnel	Adjust entrance/exit
	Left handrail	Add farside final approach	Adjust left/right handrails to new width
	Entrance funnel	Add far recognition	Add far recognition
	Final approach	Add guides or TCPs	Add farside guides or TCPs

MOVEMENT THROUGH THE LANE

3-32. Standard lane-marking patterns help the commander in two critical aspects when moving a unit through a lane:

- Lane markers help the force posture its formation for efficient passage through the defile that is caused by the lane through the obstacle.
- The combination of lane markers and TCPs gives the commander greater flexibility in the forward and rearward movement of forces.

3-33. Far-recognition and final-approach markers in a standard lane-marking pattern help commanders smoothly change assault and follow-on forces into a column formation before passing through the lane. In most cases, the situation should allow the unit to move in a column formation from the outset; however, the enemy situation and the terrain may demand that forces remain dispersed as long as possible. Far-recognition, final-approach, and entrance-funnel markers are used to trigger commanders to alter their formations.

Figure 3-4 shows how formations at various levels may change to efficiently pass through lanes. A battalion TF initially moves from its assault position in a box, V, or wedge formation depending on the number of lanes being used.

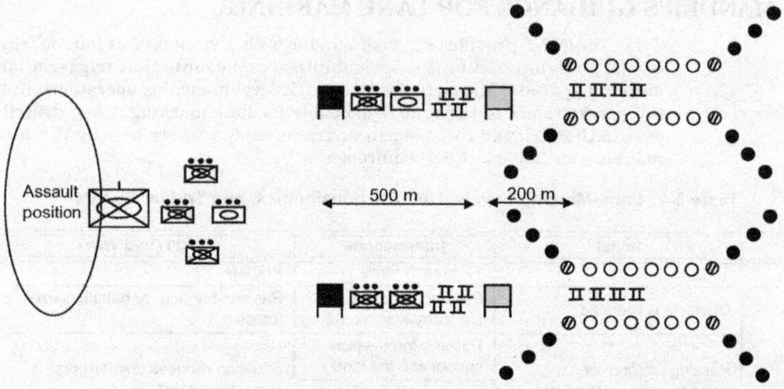

	Move out of assault position	Visually identify far-recognition marker	Pass far-recognition marker	Pass final-approach marker	Pass through exit-funnel markers
Battalion	Box (2 lanes) V (3 lanes) Wedge (3 lanes)	Box (2 lanes) V (3 lanes) Wedge (3 lanes)			
Company	Any combat formation	Column with dispersal	Column	Column	Resume formation
Platoon	Any combat formation	Any combat formation	Column with dispersal	Column	

Figure 3-4. Formation Changes Based on Lane Markers

3-34. Initially, company teams and below may move in any combat formation. Once lead elements of the battalion visually identify the far-recognition marker, company teams begin moving into a column formation with platoons still deployed. Platoons—

- Change into a column formation with dispersal as they move from the far-recognition marker to the final-approach marker.
- Change into a column formation as they move from the final-approach marker to entrance markers, using the entrance-funnel markers as guides.
- Redeploy once they pass through the exit or the exit funnels.

3-35. As the operation progresses, commanders should upgrade lane marking and establish traffic control as early as METT-TC allows. The combination of upgrading lanes from the initial marking pattern and establishing TCPs gives the commander greater flexibility. Normally, TCPs are established at forward and return far-recognition markers. Upgrading to full lane marking that is capable of supporting uninterrupted two-way traffic is generally determined by the need to pass increased volumes of traffic.

3-36. *Figure 3-5*, page 3-14, shows the flexibility that full lane marking and TCPs provide. In this example, friendly forces have created two lanes, 200 meters apart. Lane 1 is a full lane, and lane 2 is an intermediate lane. A far-recognition marker on the nearside and the farside can feed traffic to both lanes. Full lane marking and TCPs allow the commander to continue pushing combat power forward on lanes 1 and 2 simultaneously. At the same time, the farside TCP directs all return traffic to lane 1. Commanders should ensure that recovery vehicles are on site to remove immobile and destroyed vehicles.

LANE MAINTENANCE AND RESPONSIBILITIES OF TRAFFIC THROUGH LANES

3-37. As discussed in *Chapter 2*, units must plan for lane maintenance (widen lanes, upgrade lane markings, ensure that lanes remain open, ensure C^2 of traffic through the breach area). The responsibility for lane maintenance should be delineated in unit orders to ensure alignment with the scheme of maneuver. Specifically, units must determine who is responsible for—

- Maintaining lanes in tactical and protective obstacles.
- Maintaining lanes when a change occurs.
- Controlling traffic within the breach area.

3-38. *Figures 3-6 through 3-9*, pages 3-15 and 3-16, show how a unit defines responsibilities for lane maintenance and traffic control. In this example, a TF breaching operation is being conducted. Teams A and B provide suppressive fires as the support force, Team C reduces tactical obstacles as the breach force, and Team D seizes OBJs LION and TIGER as the assault force. The brigade intends to pass an armored TF through obstacles as the first element of the follow-on forces once the TF has cleared a route through the breach area. Note that the initial-lane requirement to pass Team D (the assault force) is one vehicular lane, but the breach force breaches a second lane to pass follow-on forces (an armored TF).

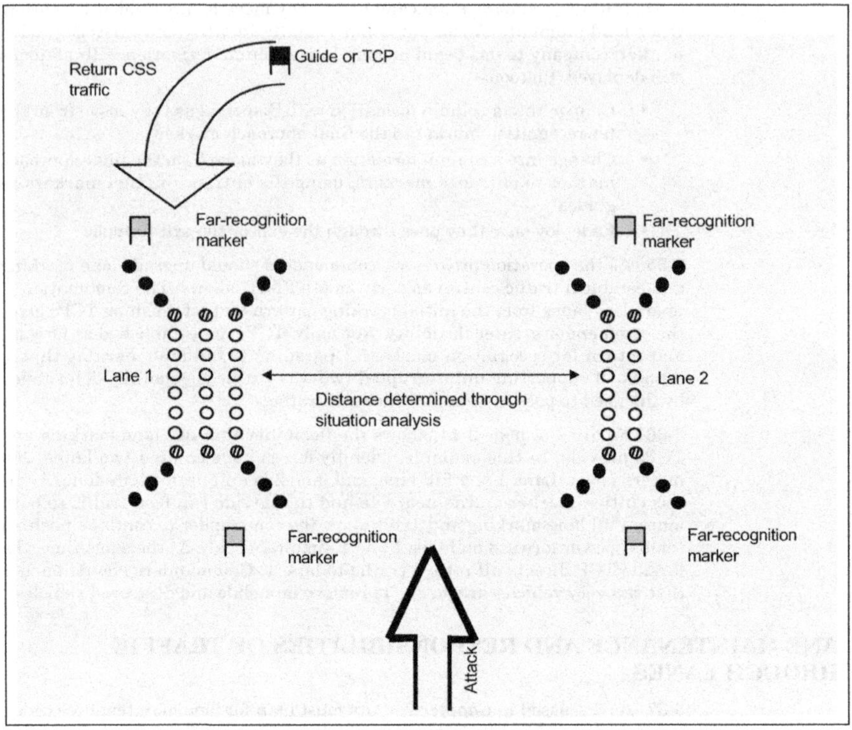

Figure 3-5. Multiple Lanes Used at Different Levels

3-39. As the breaching operation progresses, the breach force creates two lanes in the tactical obstacles. The breach force marks lanes with the initial lane-marking pattern and passes the assault force through the lanes (*Figure 3-6*). At this stage of the breaching operation, the TF maintains lanes in the tactical obstacles and controls all movement of forces within the breach area.

3-40. The assault force passes through the tactical obstacles and attacks to seize OBJs LION and TIGER. As the assault force attacks, it encounters protective obstacles and reduces and marks them. Immediately after the assault force passes through the tactical obstacles, the brigade assumes responsibility for lane maintenance in the tactical obstacles. The brigade also assumes responsibility of traffic control from the beginning of the breach area through the tactical obstacles. This includes creating additional lanes, ensuring that the lanes remain open, upgrading lane markings at the tactical obstacles, establishing TCPs, and calling units forward into the breach area. The TF is responsible for maintaining lanes through the protective obstacles (*Figure 3-7*).

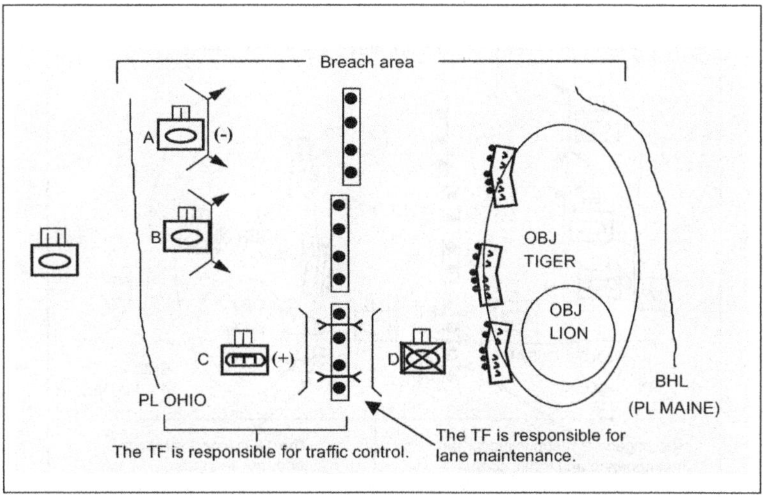

Figure 3-6. Assault Force Passes Through Tactical Obstacles

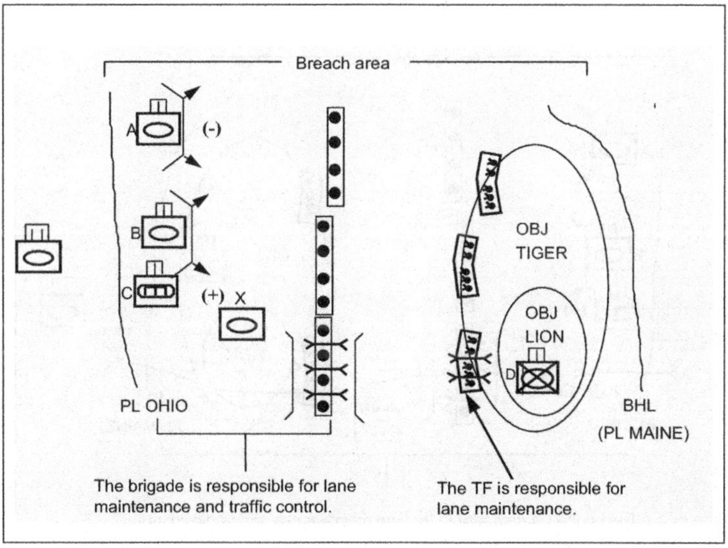

Figure 3-7. Brigade Assumes Control of Lanes in Tactical Obstacles

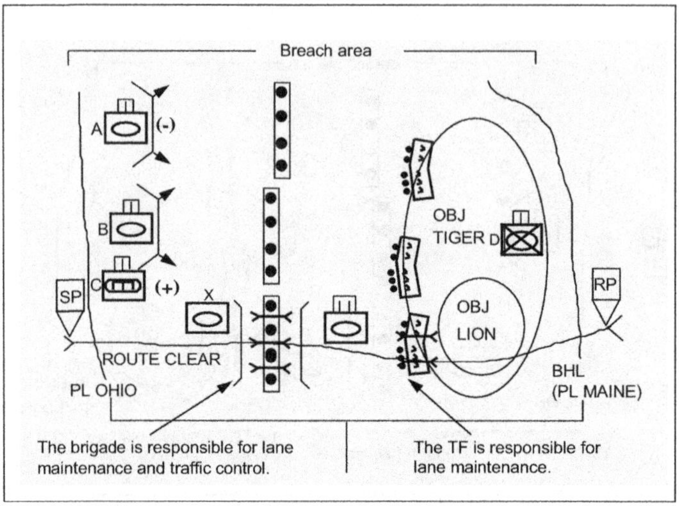

Figure 3-8. Follow-On Forces Begin to Pass Through the Breach Area

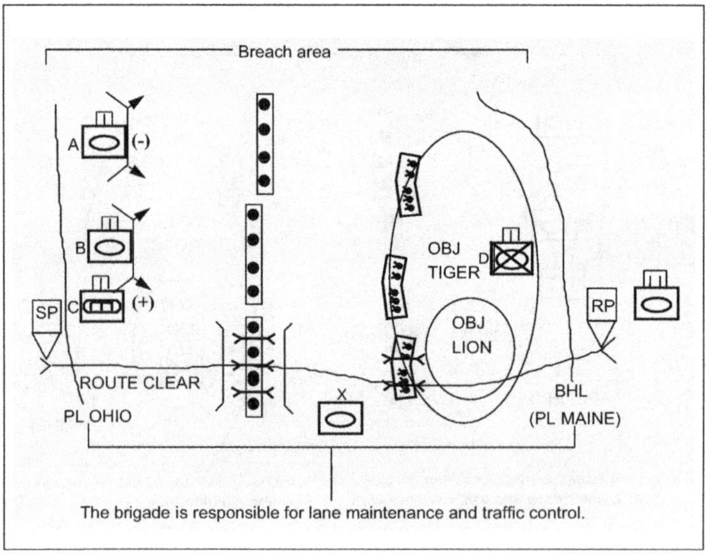

Figure 3-9. Brigade Assumes Full Control of the Breach Area

3-41. Once the assault force has seized its objective and reported that it has established a clear route up to the BHL (PL MAINE), the brigade begins to pass the follow-on TF through the breach area. At this point, the brigade controls traffic throughout the breach area. The TF is responsible for maintaining lanes in the protective obstacles (*Figure 3-8*).

3-42. As the follow-on forces cross the BHL, the breaching operation is complete; however, the obstacles within the breach area remain a constriction for other follow-on forces. At this point, the brigade maintains lanes and controls traffic. The brigade begins to focus on establishing MSRs through the breach area using corps engineer assets, which are task-organized to the brigade from the division (*Figure 3-9*).

LANE-MARKING DEVICES

3-43. Currently, the Army has no standard lane-marking device. The current marking systems in the Army's supply system are—

- Hand-Emplaced Minefield Marking System (HEMMS), national stock number (NSN) 9905-01-019-0140.
- Minefield Marking Set Number 2, NSN 9905-00-375-9180.
- Cleared-Lane Marking System (CLAMS), NSN 2590-01-205-3082.

3-44. Most lane marking in the field is done with nonstandard marking devices. Before adopting a nonstandard marking device, commanders should consider the guidelines summarized in *Table 3-2*. Coordination must occur between forces who are tasked to mark lanes and forces who are expected to use lanes to ensure that the marking material will work. For example, if a light force plans to create and mark vehicle lanes with an infrared light, vehicle drivers with enhanced (thermal) night-vision devices may not be able to see the markings.

Table 3-2. Guidelines for Lane-Marking Devices

Markers	Mounted Forces	Dismounted Forces
Handrail and funnel	Visible by TC and driver (buttoned up) from 50 meters	Visible by dismounted soldier in prone position from 15 meters
	Quick and easy to emplace (minimizes need to expose soldiers outside the emplacement vehicle)	Lightweight, quick and easy to emplace (dismounted soldier should be able to carry enough markers for the lane and still be able to fire and maneuver)
Entrance and exit	Visible by TC (buttoned up) from 100 meters	Visible by dismounted soldier from 50 meters
	Visually different from handrail and funnel markers	Visually different from handrail and funnel markers
	Quick and easy to emplace (may require soldier to dismount to emplace)	Lightweight, quick and easy to emplace
	Easily man-portable	
Final-approach and far-recognition	Visible by TC (not buttoned up) from 500 meters	Visible by dismounted soldier (on march) from 100 meters
	Visually different from each other	Visually different from each other
	Visually alterable to facilitate traffic control through multiple lanes	Visually alterable to facilitate traffic control through multiple lanes

3-45. *Figure 3-10* shows some of the devices that units currently use; they are easily procured or fabricated. This is by no means a complete listing; rather, it is intended to show commanders some of the options. Some of the general requirements for lane marking are—

- Markers must be able to withstand the rigors of the terrain, the weather, and the battle space.
- Markers should be easy to modify when visibility is limited, with minimal addition of manpower and equipment.
- Enhancements for limited visibility should be a constant source rather than a pulsating strobe. Strobes do not make the marking pattern readily apparent, particularly when approaching from an angle.

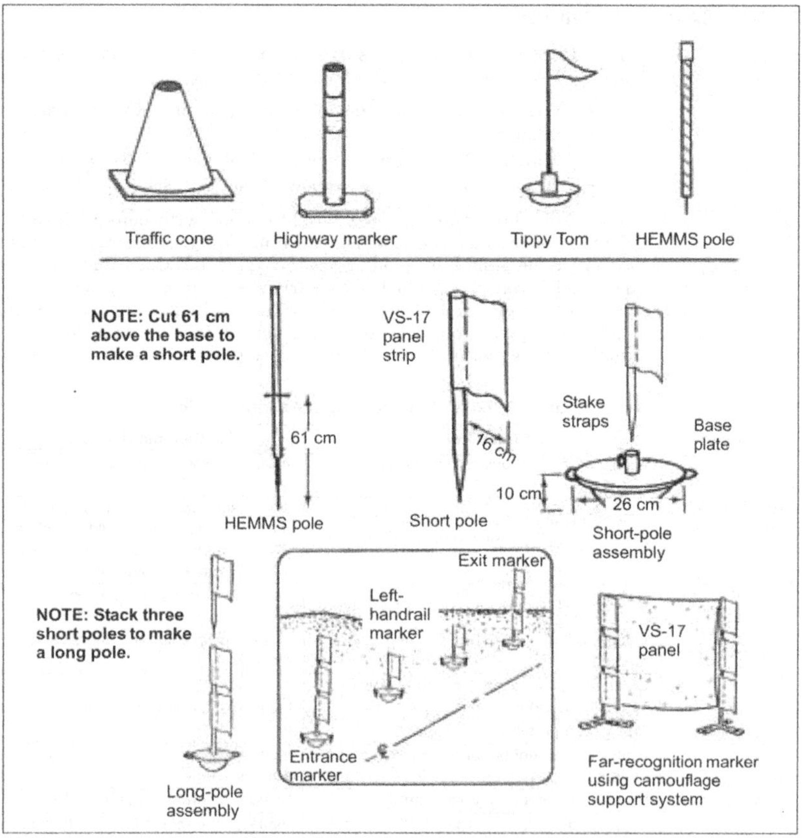

Figure 3-10. Examples of Nonstandard Marking Devices

BYPASS MARKING

3-46. Bypasses are marked using the same lane-marking patterns, marking devices, and visual cues as a lane. However, marking a bypass around an obstacle is much different from marking a lane through an obstacle because the amount of room available for forces to maneuver must be determined. While lanes through an obstacle are normally expanded to the left, bypasses are expanded away from the obstacle.

3-47. Bypasses must show where it is safe to rapidly pass around an obstacle, without unnecessarily restricting the force's ability to maneuver (*Figure 3-11,* page 3-20). To accomplish this, the marking unit will mark the bypass with a bypass-marking pattern (modified initial lane-marking pattern with right-handrail markers added). The unit will recon and initially mark the bypass 500 meters wide to allow platoons to maneuver around the obstacle. The width of the bypass is increased as time allows further recon. Limited visibility or additional obstacles discovered at the flank of the obstacle that is being bypassed may constrain the marked width of the path to less than 500 meters. (The right-handrail markers must be visible from the left-handrail markers.) In addition to bypass marking, units should consider establishing mobile guards on the friendly side of the obstacle to prevent vehicles/ personnel from entering it. All units on the battle space should possess marking material and understand how to emplace the bypass-marking pattern.

MARKING REQUIREMENTS UNDER THE NORTH ATLANTIC TREATY ORGANIZATION

3-48. The following paragraphs paraphrase the breach lane-marking requirements outlined in North Atlantic Treaty Organization (NATO) STANAG 2036. They also establish procedures used by US forces to modify intermediate and full lane-marking patterns to STANAG standard. The trend in modern warfare is increasing toward combined operations, so commanders need to be aware of their responsibilities under the NATO standard for marking hazardous areas, particularly lanes through obstacles.

COMMANDERS' RESPONSIBILITIES

3-49. STANAG 2036 outlines the minimum requirements for the lane-marking pattern before troops of other nations use the lane. Commanders must plan to convert the lane to the NATO marking standard as early as possible. When converting to the NATO standard, STANAG 2036 directs commanders to use the lane-marking devices as stated below. Within an offensive operation, this does not normally occur until the lane is matured to a full lane.

MARKING PATTERNS AND DEVICES

3-50. The intermediate lane-marking pattern satisfies the minimum lane-marking pattern that must be used before forces from another country pass through a lane. STANAG 2036 states that the entrance point, the exit point, and left and right handrails are the minimum required lane signatures.

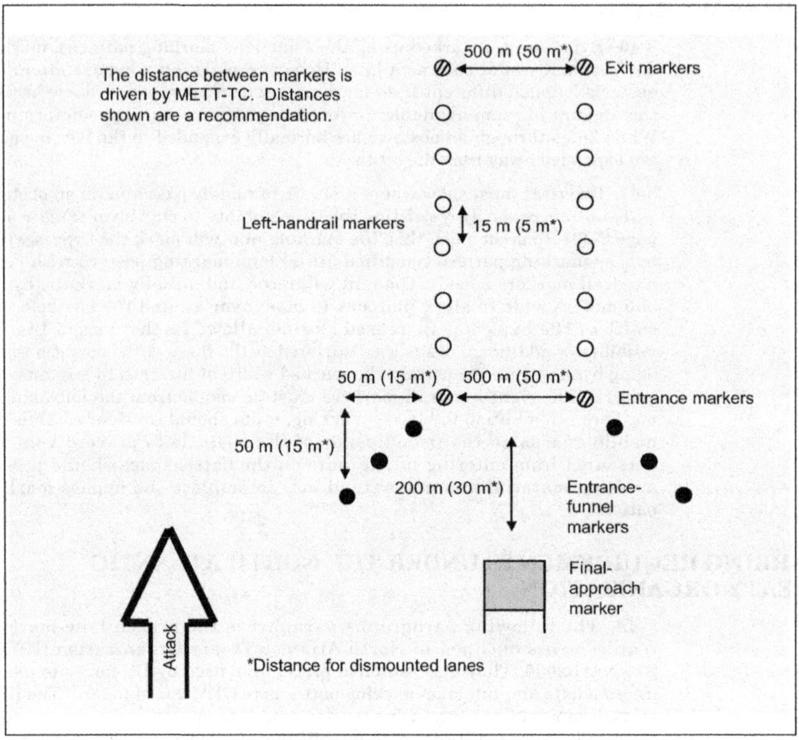

Figure 3-11. Bypass-Marking Pattern

Furthermore, STANAG 2036 requires that the entrance and exit of a lane be distinguishably different from other markers and that handrail markers be placed at intervals no greater than 30 meters. These requirements are met once lane marking is upgraded to the intermediate or full lane-marking pattern. Once the lane is marked to the the intermediate level, commanders can use the lane for allied forces to pass through without any additional marking if proper marking material is used.

3-51. *Figure 3-12* shows a NATO standard marker. The marker is placed at a right angle to the direction of travel so that the white portion of the arrow points inward to the lane, indicating the safe side on which to pass. The red portion points outward, indicating the lane limit or dangerous side of the lane. Units may fabricate NATO markers if they cannot get them through the supply system. STANAG 2036 requires that markers be visible from 50 meters under all conditions.

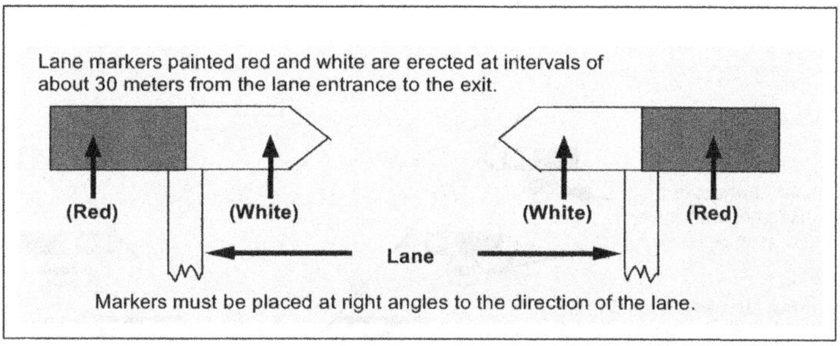

Figure 3-12. NATO Standard Marker

3-52. To convert intermediate lane marking to the NATO standard, affix NATO markers to long pickets and replace the existing entrance, exit, funnel, and handrail markers one for one. *Figure 3-13*, page 3-22, shows how NATO markers are used to convert an existing intermediate lane-marking pattern. Two NATO markers are used for entrance and exit markers to make them distinctly different. One NATO marker is affixed to each funnel marker and to each left- and right-handrail marker.

3-53. NATO uses white or green lights to illuminate markers at night. Entrance and exit markers are marked with two green lights or two white lights, placed horizontally, so that safe and dangerous markings on them are clearly visible. One white light or one green light is used on funnel and handrail markers. Commanders decide whether the light is placed on top of the NATO marker or placed so that it illuminates the marker (*Figure 3-14*, page 3-23). Lights must be visible from 50 meters under all conditions. Light sources for nighttime markers must have a continuous life of 12 hours.

COMPLEX OBSTACLES

3-54. When marking lanes through a complex obstacle, units can mark each obstacle independently or mark the entire obstacle system at once. The depth of the obstacle system, the distance between the obstacles, and the amount of marking material available determine which method is employed. See *Figures 3-15* and *3-16*, page 3-23, for examples of how complex obstacles may be marked; the examples use the initial lane-marking pattern.

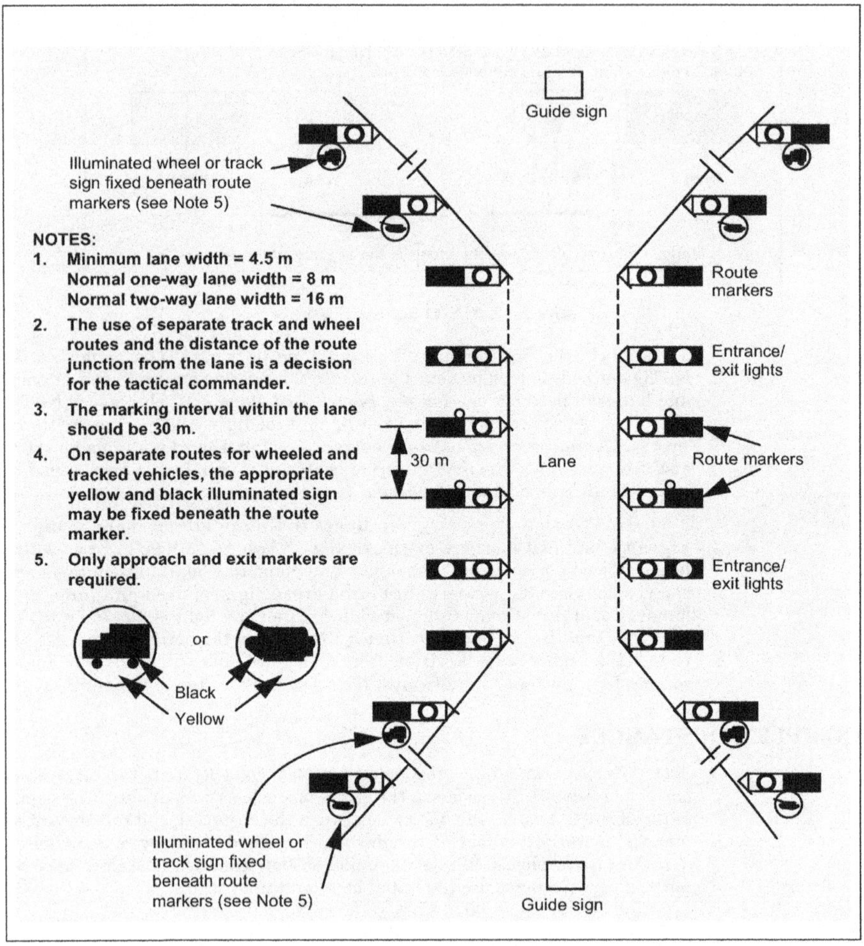

Figure 3-13. NATO Lane-Marking Standard

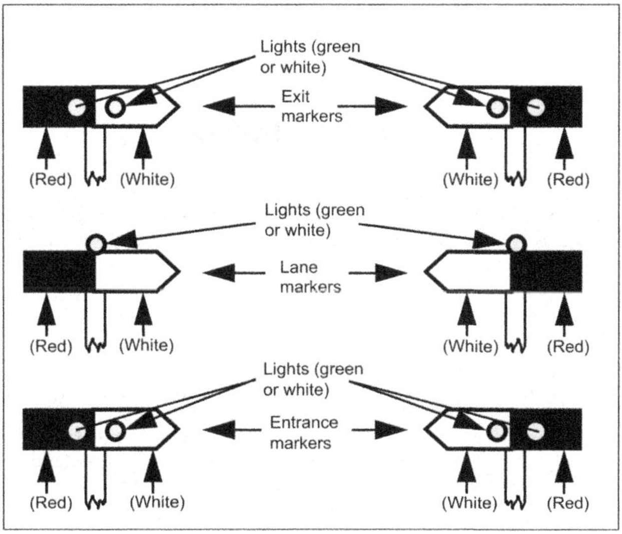

Figure 3-14. NATO Standard for Marking Lanes at Night

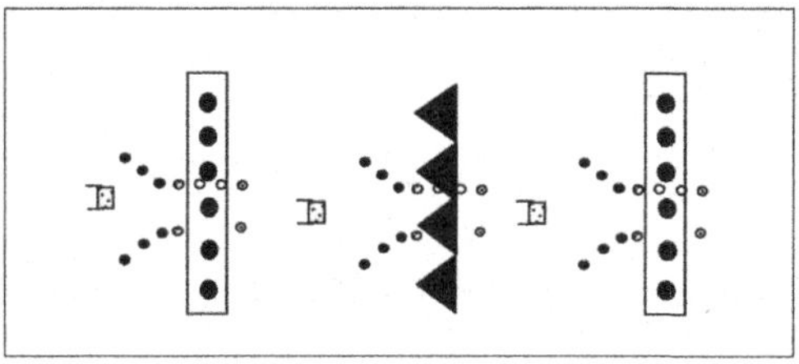

Figure 3-15. Marking a Complex Obstacle (Each Obstacle Independently Marked)

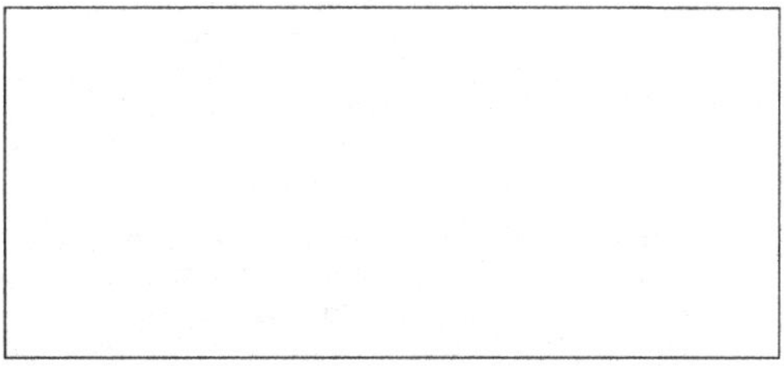

Figure 3-16. Marking a Complex Obstacle System

Chapter 4

Breaching Rehearsals

FM 101-5 gives a complete discussion of rehearsals, including types, techniques, and responsibilities. This chapter specifically discusses breaching rehearsals and considerations when conducting breaching rehearsals.

PLANNING

4-1. Rehearsals do not just happen; they must be well-planned. Successful rehearsals minimize the cost of time and maximize the understanding of the plan. Units may use a WO to establish responsibilities for preparing the rehearsal site. The WO can also identify battle drills that subordinate units should rehearse to prepare for the upcoming breaching operation. Units should include a complete rehearsal plan with an operation order (OPORD). The rehearsal plan may include a separate set of graphics if necessary.

TYPES

4-2. The five types of rehearsals are—

- Confirmation brief.
- Back brief.
- Combined-arms rehearsal.
- Support rehearsal.
- Battle drill or SOP rehearsal.

The combined-arms rehearsal and the battle drill or SOP rehearsal are most commonly used for breaching operations.

COMBINED-ARMS REHEARSAL

4-3. The maneuver unit HQ normally conducts the combined-arms rehearsal, and it is performed after subordinate units have issued their OPORD. The rehearsal ensures that subordinate plans are synchronized with those of other units and that the plans achieve the intent of the higher commander. A combined-arms rehearsal is particularly important when preparing for a complex breaching operation. It is very difficult to synchronize the actions of all the units involved in the operation.

4-4. Units should rehearse each phase of the operation, from the movement to and occupation of SBF positions through the passage of follow-on forces. The rehearsal must ensure that each unit knows how its tasks are integrated into the overall plan, in time and in space. For example, the support force must understand how long the breach force expects the reduction effort to take, so that the support force can ensure that its plan to suppress the enemy is adequate. At the same time, the breach force must understand its relative position to the support force when it is at the obstacle location. This ensures that the breach force, while moving to and reducing the obstacle, does not unnecessarily mask the fires of the support force.

4-5. At a minimum, the following should be addressed during a combined-arms rehearsal:

- The latest information concerning the enemy obstacle system.
- Actions that the higher HQ is taking to assist the breaching operation.
- The time available for emplacing smoke and activating CFZs (include location and duration).
- The criteria for lifting/shifting direct and indirect fires.
- The commitment criteria for the breach force.
- The method used to reduce the obstacle system.
- The marking to be used. One technique is to have a marked lane near the rehearsal site. All participants should view and drive/walk through the marked lane when arriving at or departing from the rehearsal site. The marking system should also be shown at support rehearsals conducted by combat service support (CSS) elements.
- The criteria and signals for the assault force to begin moving to created lanes.

BATTLE DRILL

4-6. A battle drill or an SOP rehearsal ensures that all participants understand techniques and procedures. These rehearsals are used by all echelons; but they are used most extensively by platoons, squads, and sections. For breaching operations, a battle drill or an SOP rehearsal can include lane-marking procedures or dismounted reduction techniques. Rehearsals should be conducted often to reinforce procedures that will be used during the breaching operation.

TECHNIQUES

4-7. Rehearsal techniques are limited only by the unit's resourcefulness. Generally, six techniques are used—

- Full dress.
- Reduced force.
- Terrain model.
- Sketch map.
- Terrain map.
- Radio.

Time and resource requirements for these techniques range from extensive to minimal. As the techniques are listed above, each successive one takes a decreasing amount of time and resources to prepare and conduct. Each rehearsal technique provides a different degree of understanding to participants. *Figure 4-1* shows the rehearsal techniques relative to time, resourcing, participation, and understanding.

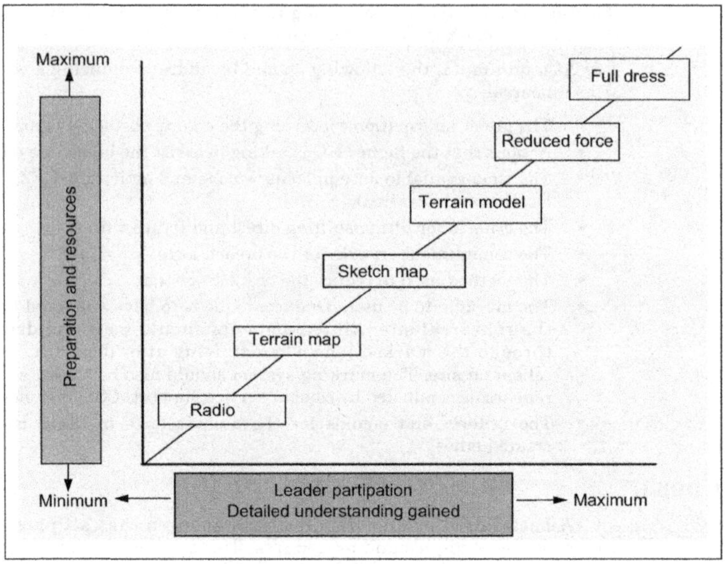

Figure 4-1. Rehearsal Techniques

4-8. When preparing for a breaching operation, use rehearsal techniques that provide the best understanding. Therefore, conduct a full-dress rehearsal if time and resources permit. Regardless of the rehearsal technique used, commanders must ensure that conditions (enemy situation, obstacle composition, visibility) are replicated in a realistic manner. For example, if a breach force is conducting a full-dress rehearsal, it should use the same obstacle system in the rehearsal as the one intelligence indicates that the enemy has emplaced. If intelligence is unavailable, the breach force should use the templated obstacle system. This requires the breach force to apply the same reduction techniques in the rehearsal that will be used during the operation.

4-9. Additionally, it is imperative that high standards be used during rehearsals. The breach force should rehearse proofing and marking obstacles. The assault force must rehearse all aspects of its assault, including the dismounted, ground fight if it is expected during the operation.

Appendix A

Metric Conversion Chart

This appendix complies with current Army directives which state that the metric system will be incorporated into all new publications. *Table A-1* is a conversion chart.

Table A-1. Metric Conversion Chart

US Units	Multiplied By	Metric Units
Feet	0.3048	Meters
Inches	2.54	Centimeters
Inches	0.0254	Meters
Inches	25.4001	Millimeters
Miles	1.6093	Kilometers
Pounds	453.59	Grams
Pounds	0.4536	Kilograms
Metric Units	**Multiplied By**	**US Units**
Centimeters	0.3937	Inches
Grams	0.03527	Ounces
Kilograms	2.2046	Pounds
Kilometers	0.62137	Miles
Meters	3.2808	Feet
Meters	39.37	Inches
Meters	1.0936	Yards
Millimeters	0.03937	Inches

Appendix B

Threat Mine Operations

This appendix is intended to complement the information presented in other manuals on threat obstacle tactics. Commanders should use this information to give added realism to unclassified staff and combined-arms team training. Obstacle employment norms can change with METT-TC factors for a given AO. Therefore, preoperational training on templating, intelligence, recon, and reduction procedures must be based on the best information available before deployment.

MINE OPERATIONS

B-1. Threat formations contain considerable organic minefield-emplacement capability. Threat rapid-mining capability presents a serious challenge to friendly maneuver.

B-2. Armored tracked mine layers known as *GMZs* (*Figure B-1*) dispense mines at a predetermined spacing of 5.5 meters. Three GMZs can lay a 1,200-meter, three-row minefield, containing 624 mines, in 26 minutes. (This does not include the 12- to 15-minute reload and travel times; travel and reload times increase during limited visibility.) Doctrinally, this minefield would be broken into several minefields, each 200 to 300 meters long. Mine-laying helicopters also support threat mine operations. The HIP and HIND-D helicopters carry two or three dispenser pods of AP or AT mines. Artillery-fired SCATMINEs can also support threat mine operations.

Figure B-1. GMZ Armored Tracked Mine Layer

B-3. Threat armies use obstacles extensively throughout the depth of their defense, and their tactics are chosen well. A threat rapidly emplaced minefield consists of three or four 200- to 300-meter rows, spaced 20 to 40 meters apart, with mines spaced 4 to 6 meters apart.

B-4. *Table B-1* provides detailed information on standard threat AT and AP minefields. Terrain and tactical situations dictate the actual dimensions and distances of minefields.

Table B-1. Normal Parameters for Threat-Style Minefields

AT Minefields	
Front (situation-dependent)	200 to 300 meters
Depth	40 to 120 meters
Number of rows	3 or 4
Distance between rows	20 to 40 meters
Distance between mines	4 to 6 meters for antitrack mines; 9 to 12 meters for antihull mines
Outlay, normal	550 to 750 antitrack mines per kilometer; 300 to 400 antihull mines per kilometer
Outlay, increased effect	1,000+ antitrack mines per kilometer; 500+ antihull mines per kilometer
Probability of destruction	57 percent for antitrack mines (750 per kilometer); 85 percent for antihull mines (400 per kilometer)
AP Minefields	
Front (situation-dependent)	30 to 300 meters
Depth	10 to 150 meters
Distance between rows	5+ meters for blast mines; 25 to 50 meters for fragmentation mines
Distance between mines	1 meter for blast mines; 50 meters (or twice the lethal radius of fragmentation) for fragmentation mines
Outlay, normal	2,000 to 3,000 HE/blast mines per kilometer; 100 to 300 fragmentation mines per kilometer
Outlay, increased effect	2 to 3 times the normal outlay
Probability of destruction	15 to 20 percent for HE/blast mines (2,000 per kilometer); 10 to 15 percent for fragmentation mines (100 per kilometer)

B-5. *Figure B-2* shows a standard rapidly emplaced minefield. The threat army typically uses such a minefield when they are in a hasty defense (offense is temporarily stalled).

B-6. *Figure B-3* shows a standard antitrack minefield.

B-7. *Figure B-4*, page B-4, shows a standard antihull minefield.

B-8. *Figure B-5*, page B-4, shows a standard AP minefield.

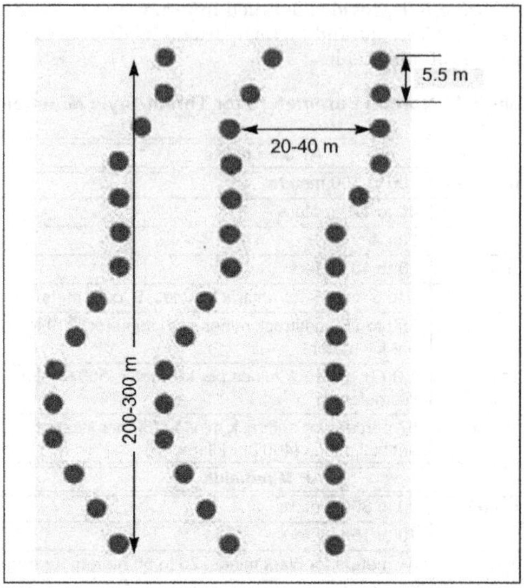

Figure B-2. Threat-Style Rapidly Emplaced Minefield

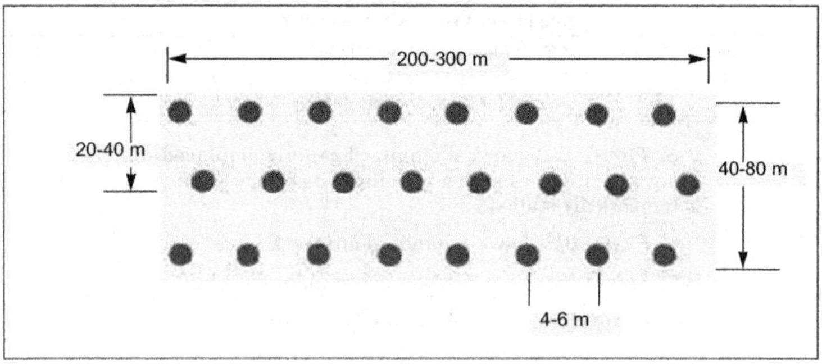

Figure B-3. Threat-Style Antitrack Minefield

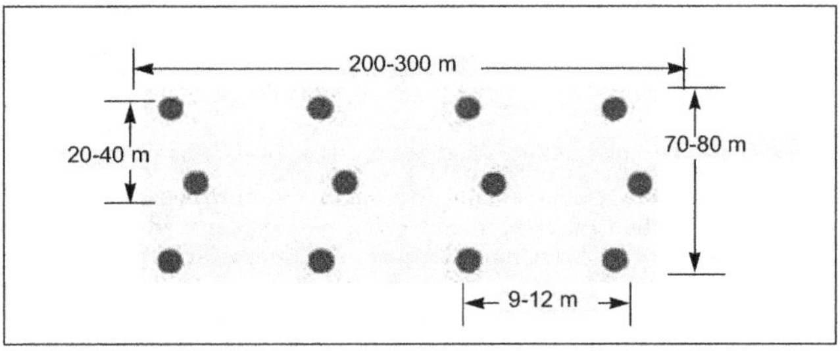

Figure B-4. Threat-Style Antihull Minefield

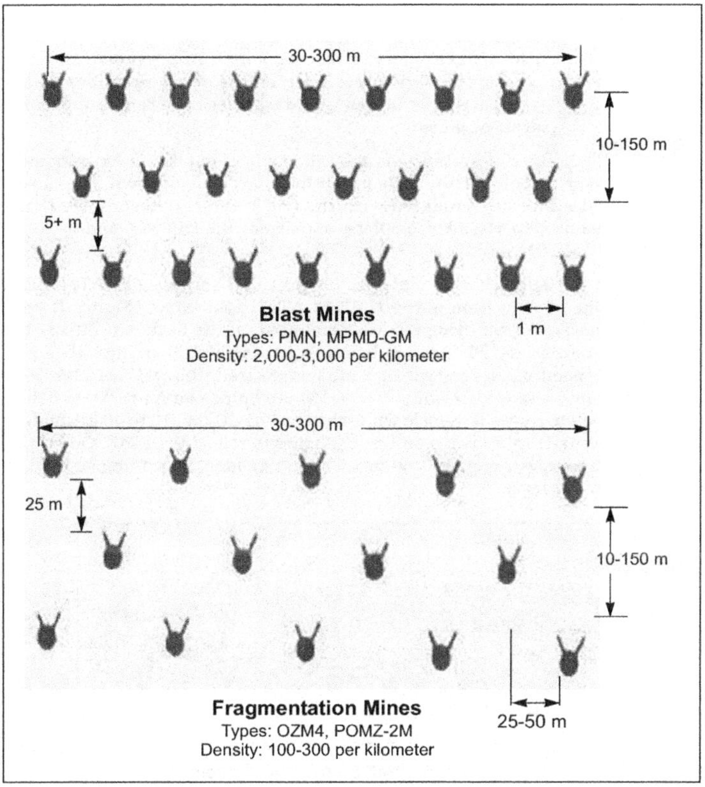

Figure B-5. Threat-Style AP Minefield

B-9. Threat armies also emplace mixed minefields. They are not the same as US mixed minefields. Threat armies normally emplace three rows of AT mines, then several rows of AP mines. AT and AP mines are not normally mixed in the same row.

B-10. Threat engineers use two fundamental drills to emplace mines:

- When emplacing armed mines, the threat uses a crew of five sappers. The first crew member (the senior man and operator) is in the mine layer's seat and monitors the operation of the mine layer and the motion of the mines in the guide chute. He also sets the mine spacing and controls the actions of the GMZ. The second and third members take the mines out of the containers and place them in the intake chute at intervals between the guide tray's drive chain. The GMZ driver steers the vehicle along the indicated route at the established speed.
- When emplacing unarmed mines, the threat uses two or three additional sappers to arm the mines. After emplacing the mines, one sapper trails the mine layer, marks emplaced mines with pennants, and partially camouflages the mines. The remaining sapper(s) then arm the mines.

B-11. Special precautions are taken when emplacing AP minefields. Threat doctrine only allows PMN mines to be surface-laid from mine layers. POMZ-2M mines are emplaced with the truck-and-tray technique. Extra effort is required to assemble, emplace, and deploy the trip wire and to camouflage the POMZ-2M mine.

B-12. Threat forces may also have ground-emplaced SCATMINE capability. One such system is the UMZ SCATMINE system (*Figure B-6*). The UMZ consists of six firing modules mounted on the back of a Zil-131 truck. Each module has 30 firing tubes, for a total of 180 firing tubes per system. Depending on the type of minefield desired, the UMZ can lay 180 to 11,520 mines without reloading. The UMZ can launch an AT or AP minefield 30 to 60 meters from the vehicle while the truck is driving 10 to 40 kilometers per hour (kph). It takes two men 1½ to 2 hours to reload the UMZ. One UMZ can lay a three-row minefield, 150 to 1,500 meters long, depending on the type of mine that is used.

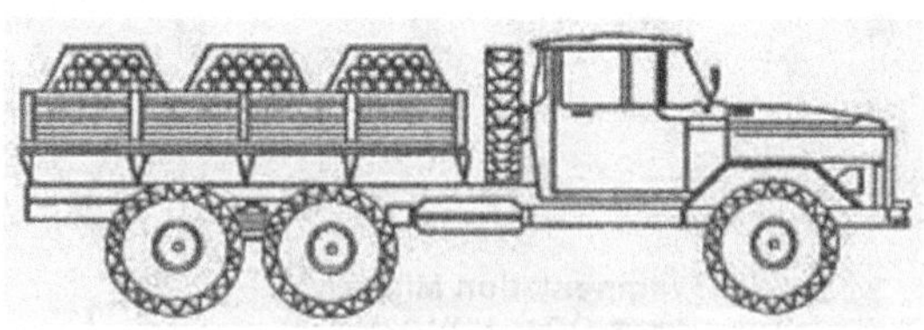

Figure B-6. UMZ SCATMINE System

B-13. For hand-emplaced SCATMINEs, there is a man-portable SCATMINE dispenser. The PKM weighs 2.63 kilograms (without the mine canister) and consists of a single launch tube with a base mount, a blasting machine, and a reel of electric ignition wire. The operator loads a propelling charge and a mine canister into the launch tube and mounts the tube on the edge of a trench or firing parapet. He then aims the tube, connects the ignition wire to the tube, and moves to a safe distance. At an initiating point, the operator connects the ignition wire to the blasting machine and initiates the system. The PKM propels the canister 30 to 100 meters, depending on the type of mine. It lays an AP minefield that is 10 by 20 meters (POM-1S or PFM-1S mine canister) or 10 by 40 meters (POM-2S mine canister). It takes a trained operator 5 minutes to set up the PKM and create a minefield. The PKM can also be used to launch the PTM-1S and PTM-3 AT mine canisters.

B-14. Threat forces use the PKM to lay minefields that protect subunit positions and flanks and the boundaries between subunits. PKM-laid minefields also cover firing lines and gaps in combat formations. The PKM can quickly close breaches in existing minefields and increase the density of mines on armor AAs.

B-15. The type and complexity of an obstacle depend on the installing unit. Maneuver and artillery soldiers usually install simple single-system minefields that are protective in nature. Engineer soldiers install complex obstacles that can include AHDs. Engineer obstacle placement is usually equipment-intensive. Threat engineer effort generally concentrates on tactical obstacles unless maneuver soldiers are unable to employ the necessary protective obstacles. Threat units continue to improve obstacles, supporting their positions by marking the friendly side of the obstacles, burying mines, and adding AHDs.

FOREIGN-MINE DATA

B-16. *Tables B-2 through B-7*, pages B-7 through B-11, provide details on foreign AT and AP mines and mine delivery systems.

Table B-2. Foreign Track-Width AT Mines

Origin	Mechanically Planted	Scatterable	Metallic	Nonmetallic	Simple Pressure	Double Impulse	Blast Hardened	AHD
Russia	TM-46 TM-57 TM-62	PTM-1	TM-46 TM-57 TM-62	TM-62B TM-62D TM-62P	TM-46 TM-57 TM-62		TM-46 TM-57	TMN-46
China	Type 72	Type 84B	Type 72	Type 72	Type 84B	Type 72	Type 72	
Yug	TMM-1		TMM-1	TMA-1 TMA-2 TMA-3 TMA-4 TMA-5	TMA-1 TMA-2 TMA-5		TMA-3 TMA-4	TMM-1 TMA-1 TMA-2 TMA-3 TMA-4 TMA-5
Czech	PT-MI-K PT-MI-BA III		PT-MI-K PT-MI-D	PT-MI-BA II PT-MI-BA III	PT-MI-BA III PT-MI-D		PT-MI-K PT-MI-BA II	PT-MI-K PT-MI-D
Italy	VS-1.6 VS-2.2 TC-.6/6 SB-81	VS-1.6 SB-81		VS-1.6 VS-2.2 TC-.6/6 SB-81			VS-1.6 VS-2.2 TC-3.6 TC-6 SB-81	VS-1.6 VS-2.2
France	M-51 M-52			M-51 M-52	M-51 M-52			M-51
UK	L9 Barmine MK-7		MK-7	L9 Barmine	L9 Barmine MK-7	L9 Barmine MK-7		L9 Barmine
Germany	PM-60 DM-11		DM-21	PM-60 DM-11	PM-60 DM-11 DM-21			PM-60 DM-11 DM-21

Table B-3. Foreign Full-Width AT Mines

Origin	Mechanically Planted	Scatterable	Tilt Rod	Magnetic Influence	Seismic Influence	Coupled	AHD
Russia	TM-62 TM-72 TM-89	PTM-3	TMK-2	TM-62 TM-72 TM-89	TM-62 TM-72 TM-89		TM-62 TM-72 TM-89
China			Type 84C	Type 84A			Type 84A
Yug	TMRP-6	KB-2	TMRP-6	TMRP-6 KB-2			TMRP-6 KB-2
Czech			PT-MI-P				
France	HPD series	Lance	M-51 M-52 Model 1954	HPD series		Model 1953	HPD series M-51 Lance
Hungary	UKB-63		UKB-63				UKB-63
Italy	VS-HCT VS-HCT-2 SB-MV	VS-SATM1 VS-HCT		VS-HCT VS-HCT-2 SB-MV VS-SATM1		Fuse option	VS-HCT VS-HCT-2 SB-MV VS-SATM1
Sweden	FFV-028			FFV-028		Fuse 15	FFV-028
UK	L9 Barmine	HB-876	L9 Barmine MK-7	L9 Barmine HB-876			L9 Barmine MK-7 HB-876
Germany	DM-31	AT-2 MIFF	AT-2	DM-31 MIFF			DM-31 AT-2 MIFF
Romania						MC-71	

Table B-4. Foreign Pressure-Fused AP Mines

Origin	Scatterable	Low Metal	Simple Pressure	Small Diameter	Low-Operating Threshold	Pronged Fuse	Blast Hardened	AHD
Russia	PFM-1		PMN PMD-6		PMN PMD-6 PFM-1	OZM series	PMN-2	
China	GLD-111	Type 72A GLD-111	Type 72A Type 58 GLD-111	GLD-111	Type 72 Type 58 GLD-111			Type 72A Type 72B
Yug		PMB-1 PMB-2 PMB-3	PMA-1 PMD-1		PMA-1 PMA-3 PMD-1	PROM-1 PROM-2 PMR-3	PMA-2 PMA-3	
Czech			PP-MI-D PP-MI-BA		PP-MI-D	PP-MI-SR PP-MI-SR II		
Italy	TS-50 VS-50 SB-33 VS-MK2	TS-50 SB-33 VS-MK2				Valmara 59 Valmara 69	TS-50 VS-50 SB-33 VS-MK2	VS-50 SB-33 VS-MK2
France		M-59		M-59	M-59		M-59	
UK	Ranger		Ranger	Ranger				
Germany		DM-11	PPM-2		DM-11		DM-11	

Table B-5. Foreign Side-Attack AT Mines

Origin	Command	Break Wire	Influence	AHD
Russia	MON-100/200/500, TM-83	LMG	TM-73/83	
Yug		TMRP-6, COBRA		
France	MIACAH F1	MIACAH F1	MIACAH F1	
UK	L14	L14	AJAX	AJAX
Germany			DM-12	
Sweden		FFV-016, AT-4		

Table B-6. Foreign Trip-Wire/Break-Wire-Fused AP Mines

Origin	Scatterable	Trip Wire	Break Wire
Russia	POM-2	POMZ-2, OZM series, MON series	OZM series, MON series
China		Type 58, Type 59, Type 66, Type 69	
Yug		PROM-1/2/KD, PMR-2/3/4, MRUD	PROM-1/2/KD, PMR-2/3/4, MRUD
Czech		PP-MI-SR, PP-MI-SR II, PP-MI-SK	
Italy	VS-SAPFM3	Valmara 59/69, BM-85, P-25, P-40, VS-DAFM1	
France		M1951/1955, MAPED F1	M1951/1955
UK		PAD	
Germany		DM-31, SM-70	

Table B-7. Foreign Mine Delivery Systems

Origin	System	Type[1]	Range	Mine Payload	Density[2]	Remarks
Russia	BM-21	R	20 km	PTM 1/3, POM-2	1.5	1,000 x 500 m per battery
	BM-22	R	35 km	PTM 1/3, POM-2	1.5	
	BKPI	F, H	NA	PTM 1/3, PFM-1, POM-2	1.5	Lays relatively narrow strips
	KSMU	F	NA	PTM 1/3, PFM-1, POM-2	1.5	
	UMZ	V	NA	PTM 1/3, POM-2	1.5	
	GMZ/PMZ-4	MP	NA	Varies	0.4 to 1.0	3 systems can emplace 1,200 m with 624 mines in 26 minutes
	VMR-1/2	H	NA	Varies	0.4 to 1.0	3 Mi-8s can lay 1,200 to 3,000 m with 1,200 mines
China	Type 83	R	15 km	Type 84	0.5	900 x 800 m per battery
Yug	M87	R	50 km	KB-2	0.5	900 x 500 m per battery

Table B-7. Foreign Mine Delivery Systems (continued)

Origin	System	Type[1]	Range	Mine Payload	Density[2]	Remarks
Italy	FIROS 25	R	22 km	VS-HCL, VS-Mk, VS-SAPFM 3	0.5	
	FIROS 30	R	35 km	VS-HCL, VS-Mk, VS-SAPFM 3	0.5	
	DAT	H	NA	Varies	1.2	Lays relatively narrow strips of up to 320 AT mines
	Istrice	V	NA	SATM1, VS-Mk 2 EL, VS-SAPFM 3	Max 1.6	Typically 360 x 140 m
	ST-AT/V	MP	NA	Varies	0.3 to 1.5	240 mines per hour
	Grillo 90	MD	NA	SATM1, VS-Mk 2 EL	0.5	5 AT or 15 AP mines per dispenser
Sweden	FFV-5821	MP	NA	FFV-028	0.4 to 1.4	300 buried mines per hour or 500 surface-laid mines per hour
France	Minotaur	V	NA	Lance	0.7	1,200 x 600 m with 800 mines
	EBG	V	NA	Lance	1.5	60 x 600 m with 40 mines
	155-mm How	A	18 km	Lance		8 mines per round
	Matenin	MP	NA	HPD series	0.4 to 1.4	500 mines per hour
UK	JP-233	F	NA	HB 876		430 mines per Tornado aircraft
	Barmine layer	MP	NA	L9 Barmine	0.4 to 0.9	Lays up to 700 mines per hour
	Ranger	V	NA	L10		1,296 mines per dispenser
Germany	Skorpion	V, H	NA	AT-2	0.4	1,500 x 50 m with 600 mines
	MW-1	F	NA	MIFF, MUSPA	0.5	55 to 500 m wide and 200 to 2,500 m
	LARS	R	14 km	AT-2	0.5	1,836 AT mines per launcher, 1,000 x 500 m per battery
	MARS	R	30 km	AT-2	0.3	1,000 x 500 m with 336 AT mines per launcher
	FFV-5821	MP	NA	DM-31	0.4 to 1.4	300 buried mines per hour or 500 surface-laid mines per hour

[1] A = artillery, H = helicopter, F = fixed-wing aircraft, R = rocket, V= vehicle dispensed, MD = man-portable dispenser, MP = mechanical planter

[2] AT mines per meter of front

Appendix C

Obstacle-Reduction Techniques

Obstacle reduction is the physical creation of a lane through an obstacle. It is a fundamental of breaching operations as discussed in earlier chapters. A number of tasks (detecting, reporting, reducing, proofing, and marking) directly support or are included in obstacle reduction.

DETECTING

C-1. Detection is the actual confirmation of the location of obstacles. It may be accomplished through recon, or it may be unintentional (such as a vehicle running into a mine). Detection is used in conjunction with intelligence-gathering operations, bypass recons, and breaching and clearing operations. There are four types of detection methods—visual, physical (probing), electronic, and mechanical.

VISUAL

C-2. Visual detection is part of all combat operations. Personnel visually inspect the terrain for the following obstacle indicators:

- Trip wires.
- Signs of road repair (such as new fill or paving, road patches, ditching, and culvert work).
- Signs placed on trees, posts, or stakes. Threat forces mark their minefields to protect their own forces.
- Dead animals.
- Damaged vehicles.
- Disturbances in previous tire tracks or tracks that stop unexplainably.
- Wires leading away from the side of the road. They may be firing wires that are partially buried.
- Odd features in the ground or patterns that are not present in nature. Plant growth may wilt or change color, rain may wash away some of the cover, the cover may sink or crack around the edges, or the material covering the mines may look like mounds of dirt.
- Civilians who may know where mines or booby traps are located in the residential area. Civilians staying away from certain places or out of certain buildings are good indications of the presence of mines or booby traps. Question civilians to determine the exact locations.
- Pieces of wood or other debris on a road. They may be indicative of pressure or pressure-release firing devices (FDs). These devices may be on the surface or partially buried.

- Patterns of objects that could be used as a sighting line. An enemy can use mines that are fired by command, so road shoulders and areas close to the objects should be searched.
- Berms. They may indicate the presence of an AT ditch.

PHYSICAL

C-3. Physical detection (probing) is very time-consuming and is used primarily for mine-clearing operations and self-extraction. Detection of mines by visual or electronic methods should be confirmed by probing. Use the following procedures and techniques when probing for mines:

- Roll up your sleeves and remove your jewelry to increase sensitivity. Wear a Kevlar™ helmet, with the chin strap buckled, and a protective fragmentation vest.
- Stay close to the ground and move in a prone position to reduce the effects of an accidental blast. When moving into a prone position—
  - Squat down without touching your knees to the ground.
  - Scan forward up to 2 meters and to the sides up to 3 meters for mine indicators.
  - Probe the area around your feet and as far forward as possible.
  - Kneel on the ground after the area is found to be clear, and continue probing forward until you are in a prone position.
- Use sight and touch to detect trip wires, fuses, and pressure prongs.
- Use a slender, nonmetallic object as a probe.
- Probe every 5 centimeters across a 1-meter front.
- Push the probe into the ground gently at an angle that is less than 45 degrees.

> **DANGER**
> **Use extreme caution when probing. If the probe is pushed straight down, its tip may detonate a pressure fuse.**

- Apply just enough pressure on the probe to sink it slowly into the ground.
- Pick the soil away carefully with the tip of the probe. Remove the loose dirt by hand if the probe encounters resistance and does not go into the ground freely. Take care to prevent detonating the mine.
- Stop probing and use two fingers from each hand to carefully remove the surrounding soil when a solid object is touched and identify the object.
- Remove enough soil to show the mine type (if the object is a mine) and mark its location. Do not attempt to remove or disarm the mine. Use explosives to destroy detected mines in place or use a grappling hook and rope to cause mines to self-detonate. Do not use metal grappling hooks on magnetic-fused mines.

C-4. Probing is extremely stressful and tedious. The senior leader must set a limit to the time a prober will actually probe. To determine a reasonable time, the leader must consider METT-TC factors, weather conditions, the threat level, the unit's stress level, and the prober's fatigue level and state of mind. As a rule, 20 to 30 minutes is the maximum amount of time that an individual can probe effectively.

ELECTRONIC

C-5. Electronic detection is effective for locating mines, but this method is time-consuming and exposes personnel to enemy fire. In addition, suspected mines must be confirmed by probing.

C-6. The AN/PSS-12 mine detector (*Figure C-1*) can only detect metal, but most mines have metal components in their design. The detector can locate and identify plastic or wooden mines by a slight metallic signature. Employment and operation procedures for the AN/PSS-12 are discussed in *FM 20-32*, and technical data is available in *Technical Manual (TM) 5-6665-298-10*. The detector is handheld and identifies suspected mines by an audio signal in the headphones.

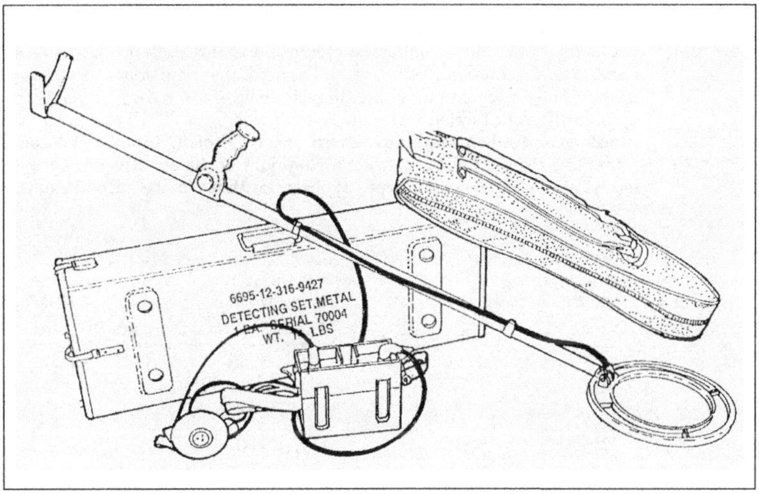

Figure C-1. AN/PSS-12 Mine Detector

C-7. As in probing, consideration must be taken for the maximum amount of time an individual can operate the detector. As a rule, 20 to 30 minutes is the maximum amount of time an individual can use the detector effectively.

MECHANICAL

C-8. The track-width mine roller is a mechanical minefield-detection system. It is most effectively deployed to lead columns on route movement, but it can be used to precede tactical formations. In column movement, unit vehicles travel a narrow path and one or two mine rollers can effectively detect mines in the path. Mine rollers can also be used to detect minefields in front of deployed tactical formations, although more than one roller is required for a good probability of detection.

REPORTING

C-9. Intelligence concerning enemy minefields is reported by the fastest means available. Spot reports (SPOTREPs) are the tactical commander's most common source of minefield intelligence. They originate from patrols that have been sent on specific minefield recon missions or from units that have discovered mine information in the course of their normal operations. The information is transmitted to higher HQ.

REDUCING

C-10. Obstacle-reduction and clearing equipment is broken down into explosive, mechanical, electronic, and manual. Combat engineers and the operators of reduction assets practice and become proficient in these reduction means. They integrate them into breach drills of units they support. The team applies different tactics, techniques, and procedures (TTP) to breach drills and prepares and rehearses them as part of the TF plan. *Tables C-1 through C-4*, pages C-5 through C-11, provide a summary of the effectiveness of reduction assets against the various types of obstacles that may be encountered.

Table C-1. Mechanical Reduction Assets Versus Threat Obstacles

Asset	Track-Width AT Mine	Full-Width AT Mine	Side-Atk AT Mine	Cmd-Det Blast Mine	AT Ditch and Other Gaps	Wire Obstacles	Remarks
IDA	Is more than 90 percent effective to 30.4 cm deep at speeds less than 10 kph (soil dependent)	Is effective against magnetic-influence mines; is more than 99 percent effective against tilt-rod mines	Do not use	Do not use	May be effective at filling in some shallow gaps or trenches	Do not use	The MCB provides almost 100 percent clearance of simple pressure-fused AT mines when used with the MICLIC. The MCB has minimal impact on the M1's maneuverability and speed when not plowing. The main gun should be traversed to the side to prevent damage in case the mine detonates when plowing. The point of breach should be relatively flat and free of large rocks and other obstructions. Plowing operations should begin about 100 m short of the suspected minefield and continue 100 m past its suspected end (typically 2 to 2½ minutes to cover 300 to 400 m overall). Plowing operations should be performed in a straight line to keep the vehicle out of uncleared ground. Fifteen percent of pressure-fused AT-mine encounters across the MCB's width can be expected to detonate under the skid shoe, probably disabling the blade. The MCB is vulnerable to coupled mines and may be stopped by log-hurdle obstacles laid within a minefield as part of a complex obstacle. A second plow should not be used to clear the center of the lane because this will simply push mines into the area cleared by the first plow.

Table C-1. Mechanical Reduction Assets Versus Threat Obstacles (continued)

Asset	Track-Width AT Mine	Full-Width AT Mine	Side-Atk AT Mine	Cmd-Det Blast Mine	AT Ditch and Other Gaps	Wire Obstacles	Remarks
MCR with IDA	Is more than 90 percent effective to 30.4 cm deep at speeds less than 10 kph (soil dependent)	Is effective against magnetic-influence mines; is more than 99 percent effective against tilt-rod mines	Do not use	Do not use	Do not use	Is effective against barbed wire	Multiple rollers are required to provide the overlap necessary for proofing the route's entire width when the MCR is used for proofing during route-clearance operations. Vehicle speed is limited to 5 to 15 kph. The main gun should be traversed to the side to prevent damage in case the mine detonates when rolling a suspected mined area. The IDA may be damaged by detonated mines.
AVLB	NA	NA	NA	NA	Can cross up to MLC 60 (70 if improved) traffic over gaps up to 17.4 m (unprepared abutments) and 18.3 m (prepared abutments)	NA	Is carried on a launcher (modified M48A5 or M60A1 chassis) Weighs 15 tons (bridge only) Can be launched in 2 to 5 minutes
Wolverine	NA	NA	NA	NA	Can cross MLC 70 traffic over gaps up to 24 m	NA	Is carried on a launcher (modified M1-series Abrams tank chassis) Weighs 12.5 tons (bridge only) Can be launched in less than 5 minutes
ACE	NA	NA	NA	NA	Is very effective	NA	

Table C-1. Mechanical Reduction Assets Versus Threat Obstacles (continued)

Asset	Track-Width AT Mine	Full-Width AT Mine	Side-Atk AT Mine	Cmd-Det Blast Mine	AT Ditch and Other Gaps	Wire Obstacles	Remarks
MiniFlail	Is effective against simple pressure-fused AP mines Is less effective against blast-hardened AP mines	Is effective	Is ineffective	Is armored against frag-menting threats Is vulner-able to command-detonated blast mines	NA	Do not use	AT mines can destroy the MiniFlail. Commanders must weigh the use of the MiniFlail against other reduction assets when conducting minefield-clearance operations. Can clear at the rate of about 1,500 square meters per hour

Table C-2. Manual Reduction Assets Versus Threat Obstacles

Asset	Pressure-Fused AP Mine	Trip-Wire/ Break-Wire-Fused AP Mine	Influence-Fused AP Mine	Cmd-Det Blast Mine	AT Ditch and Other Gaps	Wire Obstacles	Remarks
APOBS	Effectively clears 95 percent, including blast-resistant types	Effectively cuts 99 percent of trip wires in its path	Is ineffective	Is ineffective	NA	Clears a footpath	Provides a 0.6- x 45-m footpath from a 25-m standoff Can be deployed by two men in 2 minutes Consists of two manpack units weighing about 55 lbs each
Bangalore torpedo	Effectively clears 85 percent of simple pressure-fused mines; is less effective against blast-hardened mines	Effectively cuts 99 percent of trip wires in its path	Is ineffective	Is ineffective	NA	Clears a footpath	Provides a 0.6- x 15-m footpath Can be deployed by a squad in 5 minutes Consists of ten 1.5-m-long sections weighing 13 lbs each Is ineffective against AT mines
WLGH	Is ineffective	Effectively clears 99 percent of trip wires/break wires in a single pass	Is ineffective; however, it may snag the wires between the control unit and its mines	Is ineffective; however, it may snag the wires between the observer and his mines	NA	Is somewhat effective at clearing trip-wire-fused mines integrated with wire obstacles	Has a range of 80 to 100 m when launched Can be thrown 15 to 20 m when hand thrown Must consider the lethal range/radius (up to 90 m in some cases) of fragmenting AP mines that may be encountered. Select a covered position or move back an adequate distance before pulling the grapnel. Helps increase confidence by grappling an area several times Is hampered by thick vegetation/overhead obstructions that may reduce the WLGH's range/effectiveness

Table C-2. Manual Reduction Assets Versus Threat Obstacles (continued)

Asset	Pressure-Fused AP Mine	Trip-Wire/ Break-Wire-Fused AP Mine	Influence-Fused AP Mine	Cmd-Det Blast Mine	AT Ditch and Other Gaps	Wire Obstacles	Remarks
Probe	Is effective at locating all buried mines; however, extreme caution should be used when probing for small mines (less than 7.6 cm in dia) and sensitive mines (more than 10 lb of actuating force)	Is ineffective; the soldier must rely on clearance by other means and visual detection	Do not use	Ensure that observer is neutralized before approaching	NA	Is ineffective	
Wire cutter	Is ineffective	Do not cut a slack wire until both ends have been checked if break-wire-fused mines may be present. Do not cut taut trip wires until both ends have been checked for tension-release FDs	Do not use	Ensure that observer is neutralized before approaching	NA	Is effective; sound can be minimized by wrapping the wire with a cloth before cutting it	
Gunfire	Is effective against surface-laid munitions	Is effective against surface-laid munitions	Do not use	Is effective against surface-laid munitions	NA	Is ineffective	Is used as a last resort Is a relatively slow method for clearing visually detected ordnance Ensures protection of weapons operator Works better with larger caliber weapons (up to 25 mm); incendiary ammo works best May be the only option open to artillery, MP, and transportation units for dealing with SCATMINEs in the absence of engineers and reduction assets

Table C-3. Dismounted Breaching Assets Versus Threat Obstacles

Asset	Pressure-Fused AP Mine	Trip-Wire/ Break-Wire-Fused AP Mine	Influence-Fused AP Mine	Cmd-Det Blast Mine	AT Ditch and Other Gaps	Wire Obstacles	Remarks
Demolitions	Place at least a 1-lb charge within 15.2 cm of simple pressure-fused mines Ensure that the charge is placed within 2.54 cm of blast-hardened mines	Place at least a 1-lb charge within 15.2 cm of the mine after the mine at the end of a trip wire has been located Can use elevated charges if necessary against the claymore and stake-type mines	Do not use	Ensure that observer is neutralized before approaching Can use elevated charges if necessary against claymore mines	NA	Is effective when used to make an improvised bangalore torpedo	
Blast overshoes	May be effective against very small (up to 1 oz of explosive) mines Is ineffective against mines with more than 1-oz explosive weight	Can reduce casualty severity against fragmenting mines	Can reduce casualty severity against fragmenting mines	Ensure that observer is neutralized before approaching Can reduce casualty severity	NA	Is ineffective	
IBASIC	Will protect personnel near the individual who detonates a pressure-fused AP mine	Can reduce casualty severity against fragmenting mines	Can reduce casualty severity against fragmenting mines	Ensure that observer is neutralized before approaching Can reduce casualty severity	NA	Is ineffective	

Table C-4. Electronic Detection Assets Versus Threat Obstacles

Asset	Pressure-Fused AP Mine	Trip-Wire/ Break-Wire-Fused AP Mine	Influence-Fused AP Mine	Cmd-Det Blast Mine	AT Ditch and Other Gaps	Wire Obstacles	Remarks
AN/PSS-12	Is very effective at finding metallic mines Is less effective against low-metal mines	Is effective at finding metallic trip wires and the fragmenting mines to which they are normally attached Cannot detect monofilament fishing line used for trip wires; it is undetectable by the AN/ PSS-12. Visual detection and other neutralization means should be used first	Do not use	Ensure that observer is neutralized before approaching	NA	Is ineffective	Is less effective against low-metal mines and in areas contaminated with metal fragments (artillery, mortar, small arms) or areas that have heavily mineralized soils

EXPLOSIVE

C-11. The different types of explosive minefield-reduction and clearing equipment are discussed below.

M58A4 Mine-Clearing Line Charge

C-12. The MICLIC (*Figure C-2*) is a rocket-propelled, explosive line charge. It is used to reduce minefields that contain single-impulse, pressure-activated AT mines and mechanically activated AP mines. It clears a 14- by 100-meter path. The MICLIC has a 62-meter standoff distance from the launcher to the detonation point. The MICLIC's effectiveness is limited against prong AP mines, magnetically activated mines (including some SCATMINEs), top-attack mines, side-attack mines, and mines containing multiple-impulse or delay-time fuses. It also has little effect on other obstacles, such as log and concrete barriers, antivehicular ditches, and walls. The shock effect and the psychological impact of the detonation make the MICLIC a useful weapon in a close fight or in urban areas.

Figure C-2. MICLIC

NOTE: See *Figures C-3* and *C-4* for information on the MICLIC's surface danger zone.

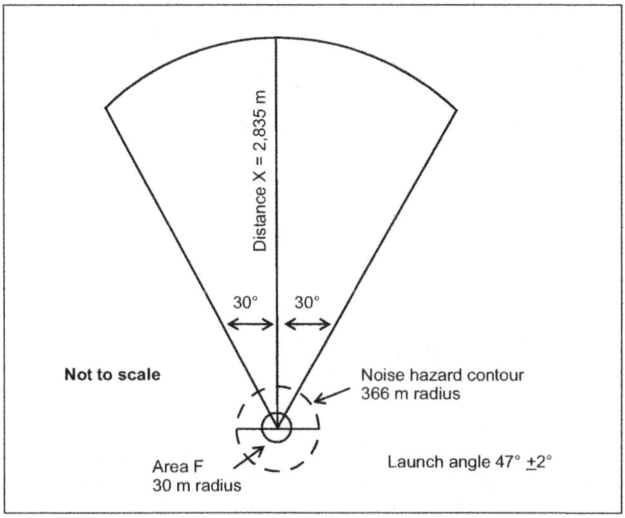

Figure C-3. Surface Danger Zone for Firing the MICLIC

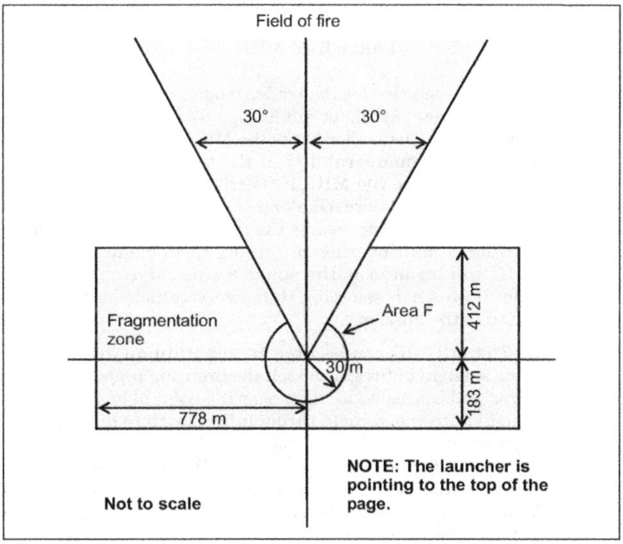

Figure C-4. Area F and Fragmentation Zone (in Detail) for Firing the MICLIC

C-13. The MICLIC is mounted on a rubber-tired trailer, or two MICLICs can be mounted on an AVLB, with the bridge downloaded using a fabricated I-beam frame. Procedures for mounting the MICLIC on the AVLB are outlined in *TM 9-1375-215-14&P*. This is called an armored vehicle-launched MICLIC (AVLM) (*Figure C-5*) and may be the preferred system because no trailer is involved to hinder the mobility of the vehicle.

Figure C-5. AVLM

C-14. Towing vehicles for the trailer-mounted MICLIC are an M113 armored personnel carrier (APC), an M9 ACE, a 5-ton wheeled vehicle, and a 2½-ton wheeled vehicle. The trailer limits the MICLIC's mobility in rough terrain and degrades the maneuverability of the towing vehicle, thereby increasing vulnerability. Since the MICLIC is critical to the breaching operation, it is moved to the point of breach along easily trafficable, covered, and concealed routes. This effectively prevents the towing vehicle from performing any other task (firing or maneuvering) or serving as an engineer squad vehicle unless MICLIC employment is the squad's only mission. This is an important consideration when selecting the towing vehicle because it must be solely dedicated to the mission.

C-15. The MICLIC can be fired from within an armored vehicle without exposing soldiers to fires, although the prefiring preparations must be done at a covered and concealed location near the point of breach. The initiating cable is brought into the vehicle through the hatch, which must be left ajar, or through the portal of the periscope, which has been removed. Therefore, the crew is not afforded NBC protection. When the MICLIC is fired from a wheeled vehicle, the crew must move to a covered position outside the backblast area. The special-purpose cable on the firing control switch is long enough to allow adequate standoff.

C-16. The vehicle operator must be proficient in all aspects of preparing and deploying the MICLIC, including the critical aspect of selecting the optimum point of breach. Although the engineer platoon leader or the breach force commander directs the operator to the point of breach, ensuring that the operator can independently accomplish the task will simplify the operation and greatly enhance its likelihood of success. The towing vehicle and the operator must be selected well in advance and be dedicated solely to the task. The operator must be included in all rehearsals and planning sessions and, if possible, during leader recons. The MICLIC crew and the breaching force must ensure that misfire/malfunction drills are addressed during the unit rehearsal.

C-17. Each MICLIC trailer transports and fires one charge, and then it must be reloaded. The AVLM can fire both MICLICs before reloading. The loaded-charge container weighs 1,283 kilograms, so a lifting device such as a 5-ton wrecker or a heavy expanded mobility tactical truck (HEMTT) is needed. Reloading, which can be done by an experienced crew in about 20 minutes, entails loading a rocket on the rail and lifting a new charge container onto the launcher. The reloading operation must be done in a covered and concealed location.

C-18. The exact limits and depth of an enemy minefield are seldom known before a breach. This is particularly true when the situation is unclear and the minefield is encountered simultaneously with enemy contact. The first and only indication that a unit is in a minefield may be when a vehicle encounters a mine. The minefield's leading edge may still be an uncertainty because the vehicle could have hit a mine in an interior row. The number of MICLICs needed to clear a single lane through a minefield—

- Less than 100 meters deep is one (see *Figure C-6*, page C-16). The minefield's leading edge is identified and, if possible, confirmed by recon. The MICLIC is deployed from a minimum standoff distance of 62 meters from the minefield's leading edge.
- More than 100 meters deep or of uncertain depth is two or more (see *Figure C-7*, page C-16). If the leading edge cannot be identified, the MICLIC is deployed 100 meters from the possible edge or stricken vehicle. When the first MICLIC is detonated, a second MICLIC moves 25 meters into the first MICLIC's path and fires its charge. This extends the lane an additional 87 meters. Additional MICLICs are used for minefields of extreme depth, and each one moves down the lane 25 meters into the path created by the previous charge.

C-19. The neutralization of mines by blast depends on the peak pressure and the impulse. For the MICLIC, the impulse is at a maximum of 3 meters from the line charge (on both sides) and decreases the closer it gets toward the line charge to a minimum of 1 meter from the line charge. This decrease on impulse causes a *skip zone* (see *Figure C-8*, page C-17). This does not mean that neutralization is equal to zero percent; it means that it is not equal to 100 percent. Mines that are buried deeper than 10 centimeters and located 1 to 2 meters from the line charge have a high probability of not being neutralized.

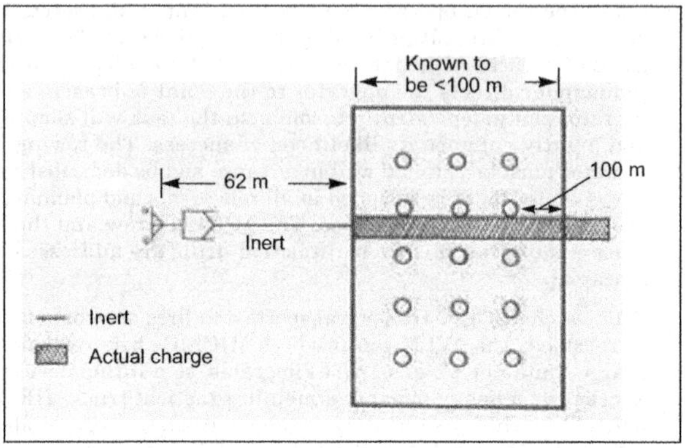

Figure C-6. MICLIC Employment in a Minefield Less Than 100 Meters Deep

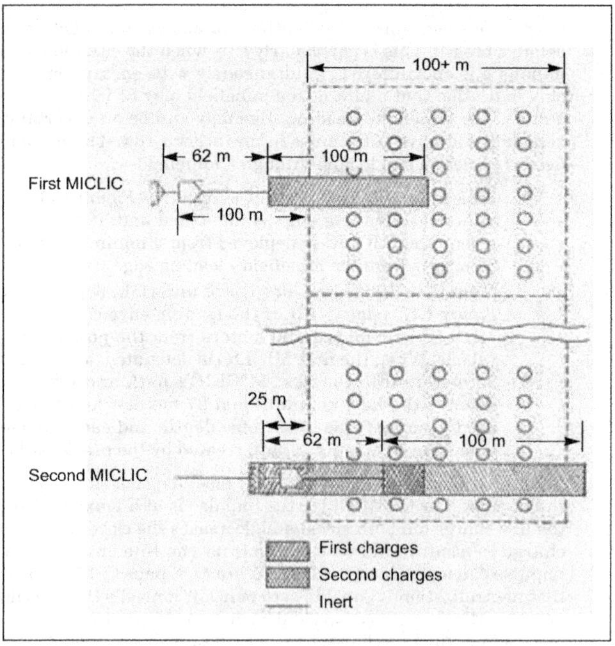

Figure C-7. MICLIC Employment in a Minefield of Uncertain Depth or Greater Than 100 Meters

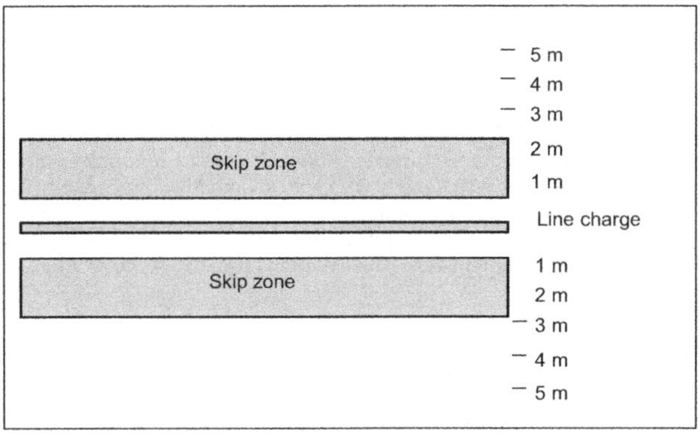

Figure C-8. MICLIC Skip Zone

Antipersonnel Obstacle Breaching System

C-20. The Antipersonnel Obstacle Breaching System (APOBS) (*Figure C-9*) is a man-portable device that is capable of quickly creating a footpath through AP mines and wire entanglements. The APOBS is normally used by combat engineers, infantry soldiers, or dismounted armored cavalry personnel. It provides a lightweight, self-contained, two-man, portable line charge that is rocket-propelled over AP obstacles from a standoff position away from the obstacle's edge.

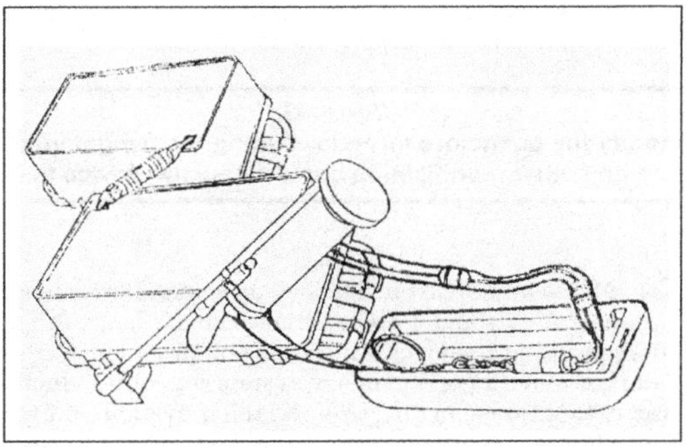

Figure C-9. APOBS

C-21. For dismounted operations, the APOBS is carried in 25-kilogram backpacks by no more than two soldiers for a maximum of 2 kilometers. One backpack assembly consists of a rocket-motor launch mechanism containing a 25-meter line-charge segment and 60 attached grenades. The other backpack assembly contains a 20-meter line-charge segment and 48 attached grenades.

The total weight of the APOBS is about 54 kilograms. It is capable of breaching a footpath about 0.6 by 45 meters and is fired from a 25-meter standoff.

M1A1/M1A2 Bangalore Torpedo

C-22. The bangalore torpedo (*Figure C-10*) is a manually emplaced, explosive-filled pipe that was designed as a wire breaching device, but it is also effective against simple pressure-activated AP mines. It is issued as a demolition kit and consists of ten 1.5-meter tubes. Each tube contains 4 kilograms of HEs and weighs 6 kilograms. The kit clears a 1- by 15-meter lane.

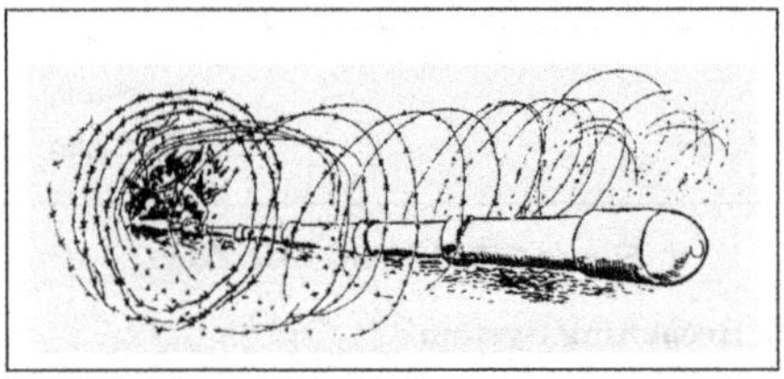

Figure C-10. Bangalore Torpedo

C-23. The bangalore torpedo is used by dismounted infantry and engineer soldiers. An individual soldier or a pair of soldiers connects the number of sections needed and pushes the torpedo through the AP minefield before priming it. A detailed recon is conducted before using the bangalore torpedo to ensure that trip wires have not been used. The bangalore torpedo generates one short impulse and is not effective against pronged, double-impulse, or pressure-resistant AP and AT mines.

WARNING
Do not modify the bangalore torpedo. Cutting the bangalore in half or performing any other modification could cause the device to explode.

MECHANICAL

C-24. MCBs and MCRs are fielded as armor battalion countermine sets that contain 12 MCBs and 4 MCRs. Blades create lanes through minefields, while rollers are used to detect minefields and proof lanes created by other means. Rollers are not a good primary system for lane reduction because multiple mine detonations may destroy the roller system and the vehicle pushing it. (The roller is designed to resist four conventional-mine [22.5 pounds HE] or three SCATMINE strikes, depending on the mine type.)

C-25. The ACE and dozer blades were not designed for reducing minefields and should only be used as a last resort. Using them is extremely time-consuming and hazardous to the crew and equipment. However, ACEs can effectively clear a lane through AP scatterable minefields because they sustain little or no damage and offer protection to the crew. When using a dozer to clear a path through a minefield, the operator is exposed to mine effects. Before clearing begins, sandbag or *up-armor* the operator's cabin and clear the lane of trip wires. When using an engineer blade to clear a path through a scatterable minefield, use the skim technique (*Figure C-11*). Start skimming 100 meters from the suspected leading edge of the minefield.

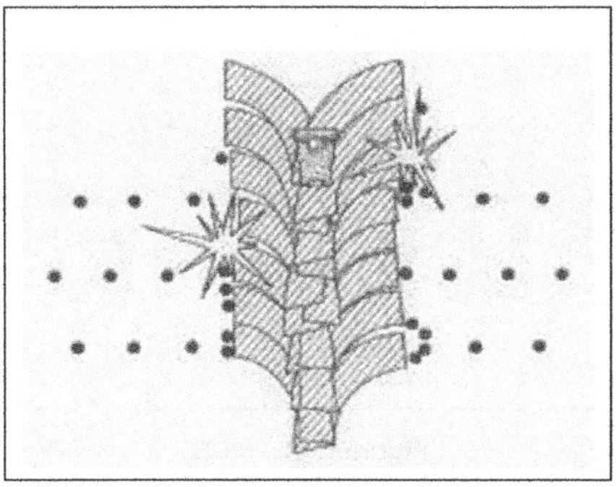

Figure C-11. Skim Technique

C-26. The ACE and dozer blades are very effective against AT ditches and gaps. Normally, during a breaching operation, dozers will not be available due to their inability to keep pace with the attacking forces. Although a single dozer or an ACE is capable of reducing an AT ditch, a technique that has been successfully used and has decreased the time to reduce an AT ditch/gap is to use two ACEs or an ACE/dozer team. When using this technique, position one ACE parallel to the AT ditch and push dirt in front of the second ACE or dozer that is oriented perpendicular to the AT ditch. Use an ACE to push dirt into the AT ditch.

Mine-Clearing Blade

C-27. The MCB (*Figure C-12*, page C-20) is used to remove land mines and create a lane in a minefield. It consists of a blade arrangement with scarifying teeth to extract mines, a moldboard to cast mines aside, and leveling skids to control the blade's depth.

C-28. The MCB lifts and pushes mines, which are surface-laid or buried up to 31 centimeters deep, to the side of track-width lanes. The blade has three depth settings—21, 25, and 31 centimeters. The blade creates a 1.5-meter cleared path in front of each track. *Figure C-13*, page C-20, shows the inside clearance distances between the tracks of common tracked vehicles in relation to the uncleared area left by the MCB.

C-29. Mines armed with AHDs, magnetic fuses, or seismic fuses may be activated when they are lifted by the blade; and they may disable the blade. Mines lifted by the blade are left in the spoil on each side of the furrowed path and remain a hazard until they are removed. Double-impulse mines that are lifted into the spoil on the side can function into the hull of the plowing vehicle.

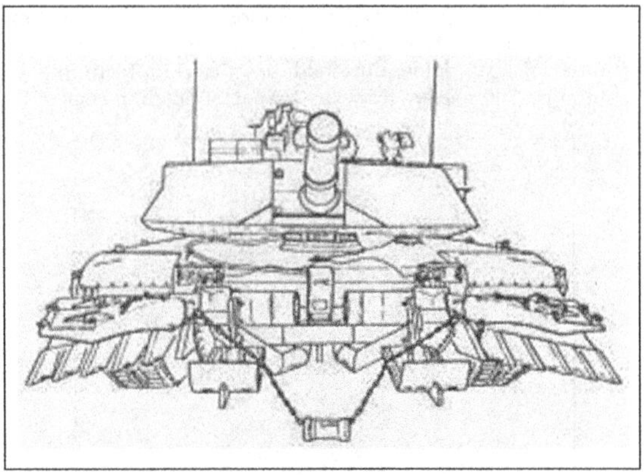

Figure C-12. MCB

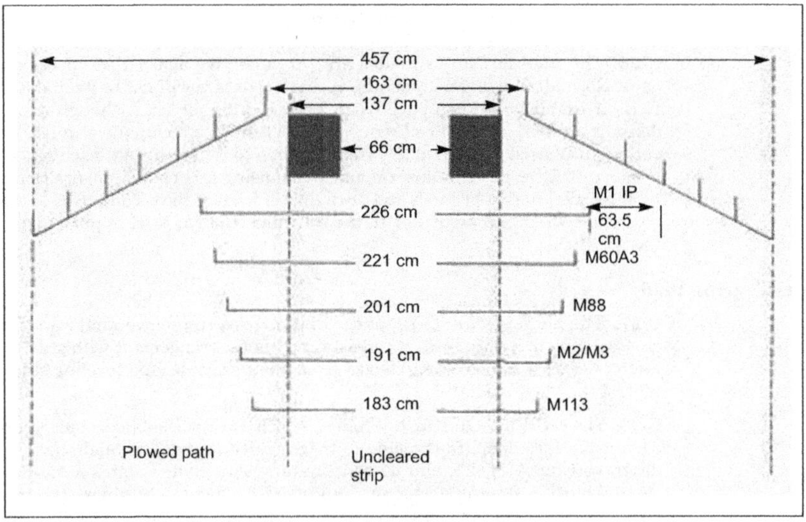

Figure C-13. MCB Width Compared to Tracked-Vehicle Widths

C-30. The skid shoe for each blade exerts adequate pressure to activate most single-impulse mines, which effectively clears a section of the centerline by explosive detonation. This action may disable the blade. Multiple-impulse pressure fuses encountered by the skid shoe are not defeated. A dog-bone assembly between the blades defeats tilt-rod mines. The improved dog-bone assembly (IDA) projects a magnetic signature and defeats tilt-rod and magnetic mines.

C-31. The MCB weighs about 3,150 kilograms and can be mounted on an M1 tank without special preparation or modification. Mounting requires lift capability and takes up to an hour, so it must be mounted well in advance of the mission. It is not easy to mount or transfer the MCB to another tank under battle-space conditions.

C-32. Once mounted, an electric motor raises and lowers the blade. When the blade is raised, it minimally effects the M1's maneuverability and speed. This will not greatly impact the use of the weapon system except when the blade is in operation. The MCB is also equipped with an emergency, quick-disconnect feature.

C-33. The M1 tank crew should maintain a speed of 8 to 10 kph while plowing, depending on soil conditions. The tank cannot maneuver but must continue in a straight path through the minefield to avoid damaging the blade. The main gun must be traversed to the side during plowing because mine detonation under the blade may cause the gun to be thrown violently into the air, damaging the tube. The area selected for the lane must be relatively flat and free of rocks or other obstacles.

C-34. The operator begins plowing about 100 meters from the estimated leading edge of the minefield. He creates a lane extending another 100 meters beyond the estimated far edge of the minefield to ensure that the lane extends through the entire minefield. Multiple vehicles using the lane will deepen the cut made by the MCB, and pressure-fused mines left in the uncleared strip will be dangerous. The uncleared strip should be cleared as soon as possible.

Mine-Clearing Roller

C-35. The MCR (*Figure C-14*, page C-22) consists of a roller assembly, a mounting kit, and a hand winch. The roller assembly weighs about 9,072 kilograms and consists of two push beams mounted to the front of the tank. The rollers are designed to defeat most single-pulse, pressure-activated AT and AP mines. The roller creates a 1.1-meter-wide cleared path in front of each track.

C-36. *Figure C-15*, page C-22, shows the inside clearance distances between the tracks of common tracked vehicles in relation to the uncleared area left by the MCR. An IDA and a chain between the rollers defeats tilt-rod mines. The IDA can be fitted to the roller. The roller is designed to withstand multiple mine explosions before damage; however, this depends on the size of the mines. Large blasts may destroy the roller or the vehicle or injure the crew.

C-37. The roller can be mounted on an M1 or M60 tank that is modified with a permanently attached mine-roller mounting kit. Mounting the roller to a tank is a cumbersome, time-consuming operation because it is very difficult

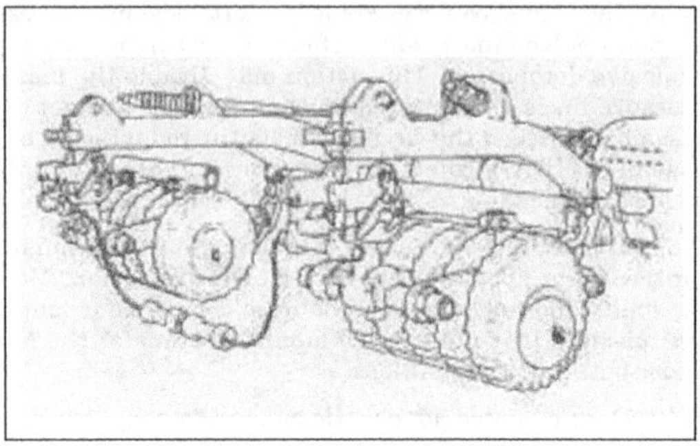

Figure C-14. MCR

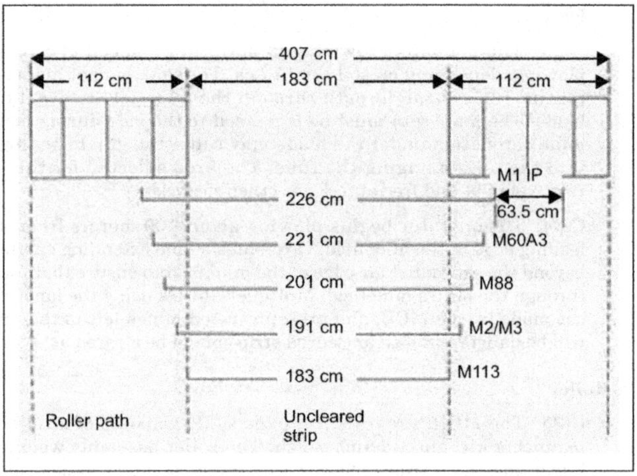

Figure C-15. MCR Width Compared to Tracked-Vehicle Width

under battle-space conditions and requires lift capability. The roller tank is limited to a speed of 5 to 15 kph. When used in a suspected minefield, the MCR must travel in a relatively straight path. Tight turns may cause the roller to deviate from the path of the track and leave the tank vulnerable to mines. Ground fluctuations, bumps, and berms may cause the roller to lift from the ground and miss mines.

C-38. Due to the roller tank's weight and width, units must use caution when crossing an AVLB. In this situation, the curbing from the bridge is removed. To prevent damage to the bridge's hydraulic line, the tank driver uses a strap to lift the IDA and the chain when crossing the bridge. The main gun must be traversed to the side when a mine encounter is possible or imminent. A mine blast can throw the roller or parts of the roller violently into the air and damage the tube. The main gun should only be fired from a temporary halt.

C-39. When the situation and the mission permit, MCRs are used with lead/ security elements of the breach force to detect the minefield's leading edge. Rollers are also used to proof lanes in obstacles that are reduced by other means, such as a MICLIC or an MCB. The roller may also be used to lead a supported element traveling in a tactical formation other than a column, but it is less effective than other methods because—

- A vehicle that is not directly behind the roller may encounter mines.
- The roller may travel well into or completely through a widely spaced minefield without encountering a mine, thus giving the formation a false sense of security.
- A mine encountered by the roller may not be on the minefield's leading edge.
- The roller vehicle is extremely vulnerable since it can only use its weapon system from a temporary halt.

C-40. If rollers are to participate in a breaching operation, they must be mounted before rehearsals. Unmounted rollers that are not being used for the mission are carried in the TF formation on M916 tractor trailers. Rollers require lift capability (such as an M88), a secure location, and 30 to 60 minutes to mount on a tank that is fitted with a mounting kit.

Panther

C-41. The M60 Panther (*Figure C-16*) is one of several developmental countermine systems that US forces used during operations Joint Endeavor and Joint TF Eagle. The Panther is a remotely controlled vehicle with mine rollers, and it is used to proof lanes and assembly areas. The system consists of a turretless M60 tank, Israeli mine rollers, an antimagnetic actuating device, and a remote control unit (RCU) that is mounted in a separate vehicle. Additionally, a remote video camera allows the operator to see the road ahead.

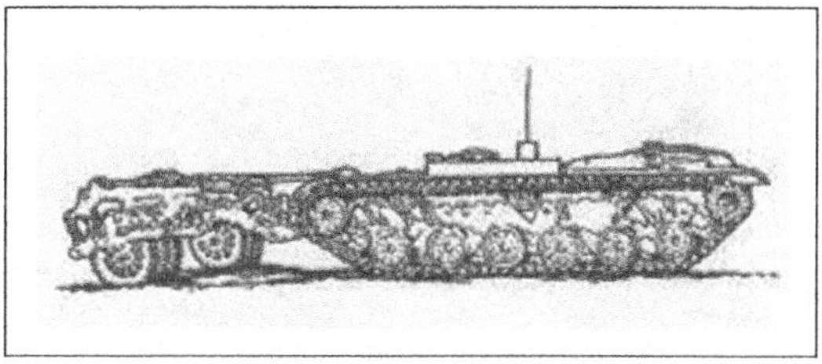

Figure C-16. Panther

C-42. During route-clearance or proofing operations, a Panther is the lead vehicle. It is followed closely by an armored control vehicle, usually an M113. The control vehicle contains the Panther operator, the RCU, and the monitor. The monitor displays the route being proofed or cleared through a camera mounted on the Panther. The Panther is controlled from the commander's cupola or troop hatch of the control vehicle. The control vehicle should be about 200 to 300 meters behind the Panther, and its hatches should be secured open. Crew members in the control vehicle should be wearing Improved Body Armor System, Individual Countermine (IBASIC) protective garments.

C-43. Mine rollers can be raised for limited travel while mounted on the Panther. If the distance is excessive, the rollers must be transported on a cargo carrier. Rollers must be adjusted before every mission to ensure that they have contact with the ground and that their weight is uniformly distributed. Conduct at least three passes to ensure proper coverage and overlap of rollers. Passes should have a minimum of 30 centimeters overlap. Inside roller distances are the same as an MCR.

M1 Grizzly

C-44. The Grizzly (*Figure C-17*) provides the capability for breaching obstacles comprised of mines, wire, posts, rubble, and tank ditches to create a lane for other vehicles to follow. The Grizzly's primary features are a full-width, 4.2-meter MCB and a power arm. The power arm has a reach of 9 meters and a bucket capacity of 1.2 cubic meters. Its primary missions are to reduce berms and fill AT ditches.

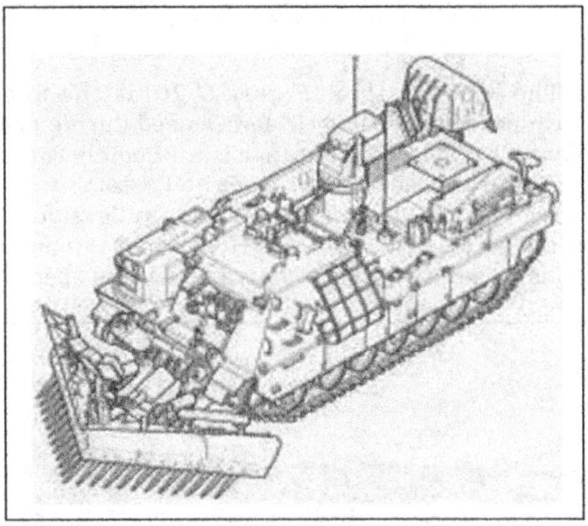

Figure C-17. Grizzly

C-45. The Grizzly lifts and pushes mines, which are surface-laid or buried up to 31 centimeters deep, to the side of full-width lanes. The blade has multiple depth settings, depending on the mission, and it creates a 4.2-meter-wide cleared path. When plowing, the Grizzly is restricted to less than 10 kph, depending on soil conditions. The operator begins plowing about 100 meters from the estimated leading edge of the minefield. He creates a lane extending another 100 meters beyond the estimated far edge of the minefield to ensure that the lane extends through the entire minefield.

C-46. The Grizzly has integrated digital features to enhance battlefield awareness. Some of the digital features are thermal and video cameras, ground-speed sensors, terrain-mapping sensors, and an integrated commander's control station.

Mine-Clearing/Armor-Protection Kit

C-47. The mine-clearing/armor-protection (MCAP) kit consists of two parts—the mine-clearing rake and the armor protection.

- **Mine-clearing rake.** The rake uses a tine that is mounted on a diagonal beam. The rake assembly is designed to sift through the soil, lift out mines, and windrow buried and surface-laid mines to the right of the vehicle. The system clears a 30-centimeter-deep path through a minefield. The rake has a skid shoe that acts as a depth control guide for the operator.
- **Armor protection.** The armor protects the crew against mine blast, small-arms fire, and artillery fire. Protection is also provided for the engine, the fuel tank, and exposed hydraulic cylinders and lines. Ballistic glass blocks are provided at each vision port to permit unrestricted view and operation of the vehicle and the equipment.

The MCAP is mounted on a D7 dozer to perform minefield reduction and lane widening. Proofing the lane must be conducted after the dozer has cleared the lane because some AP mines may still be left in the lane.

Armored Vehicle-Launched Bridge

C-48. The AVLB (*Figure C-18*) is carried on a launcher (a modified M48A5 or M60A1 chassis) and weighs 15 tons. A buttoned-up, two-man crew can launch the AVLB in 2 to 5 minutes. The AVLB can be retrieved from either end; however, one man is exposed to fire while guiding and connecting it. The AVLB is capable of carrying military load class (MLC) 60 (70 with improved bridge) tracked loads across a 17.4-meter gap with unprepared abutments or an 18.3-meter gap with prepared abutments.

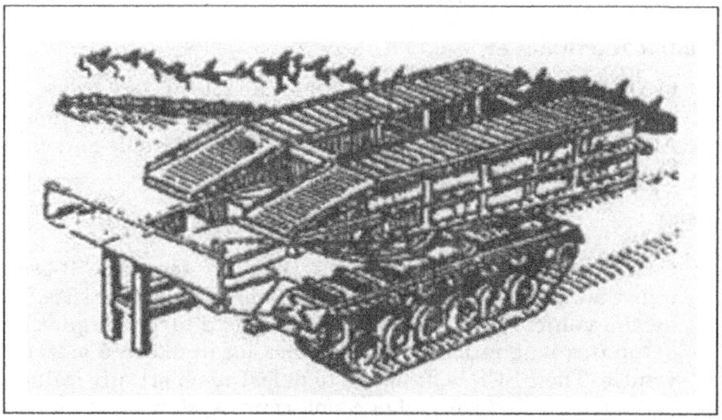

Figure C-18. AVLB

C-49. The AVLB has the following limitations:

- The scissor launch requires 10 meters of overhead clearance.
- The maximum launch slope is—
  - 28 percent uphill.
  - 19 percent downhill.
  - 11 percent on a side slope.

Wolverine

C-50. The AVLB will eventually be replaced by the Wolverine (*Figure C-19*). The bridge is carried on a launcher (a modified M1 tank chassis) and weighs 12.5 tons. It can be launched in less than 5 minutes by a buttoned-up, two-man crew and can be retrieved from either end of the bridge in less than 10 minutes. The Wolverine can cross an MLC 70 tracked vehicle over 24-meter gaps. The launcher has a maximum speed of 83 kph.

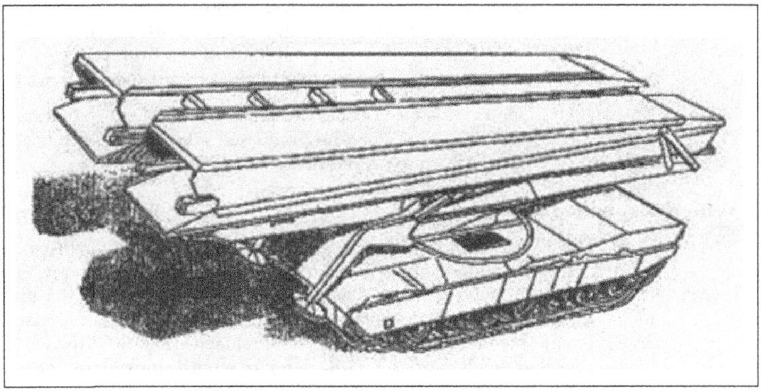

Figure C-19. Wolverine

M9 Armored Combat Earthmover

C-51. The M9 ACE (*Figure C-20*) is capable of performing mobility tasks in support of light and heavy forces. These tasks include gaps and AT ditches. Although an ACE can be used to reduce a scatterable minefield, it should only be used as a last resort.

ELECTRONIC

C-52. The Field-Expedient Countermine System (FECS) is a series of copper coils that fit over the front of tracked and wheeled vehicles. Power is supplied by the vehicle's battery. The coils emit a large magnetic signature that detonates magnetically fused mines located 2 to 5 meters in front of the vehicle. The FECS is designed to defeat magnetically influenced mines only and must be used with other countermine systems.

Figure C-20. M9 ACE

MANUAL

C-53. While mechanical and explosive reduction procedures are normally preferred, a unit may have to use manual procedures for the following reasons:

- Explosive and mechanical mobility assets are unavailable (not on hand or nonmission capable).
- Explosive and mechanical mobility assets are ineffective against the type of obstacle or because of terrain limitations (see *Table C-1*, page C-5).
- Stealth is required.

C-54. Manual procedures involve dismounted soldiers (usually engineers) using simple explosives or equipment to create a lane through an obstacle or to clear an obstacle. These procedures expose the soldier and may be manpower- and time-intensive. However, this method works in most situations and under most conditions. Certain types of terrain, weather, and sophisticated fuses can severely degrade the effectiveness of rollers, plows, and line charges.

Techniques

C-55. As discussed, manual procedures are effective against all obstacles under all conditions. Most types of obstacles do not cause casualties directly, but minefields have the potential to cause direct casualties. The enemy possesses a significant mechanical, mine-burying capability. It has the capacity and the propensity for the labor-intensive effort required to bury mines by hand; however, the enemy often lays mines on the surface. Buried mines are usually found in a highly prepared defense. When training for the reduction of surface-laid and buried minefields, always assume the presence of AHDs and trip wires until proven otherwise. Manual reduction of surface-laid and buried minefields is discussed below:

- **Surface-laid minefield.** Engineers first use grapnel hooks from covered positions to check for trip wires in the lane. The limited range of the tossed hook requires the procedure to be repeated through the estimated depth of the obstacle. A demolition team then moves through the lane. The team places a line main down the center of the lane, ties the line from the explosive into the line main, and places blocks of explosive next to surface-laid mines. After the mines are detonated, the team makes a visual check to ensure that all mines were cleared before directing a proofing roller and other traffic through the lane. Members of the demolition team are assigned special tasks, such as grappler, detonating-cord man, and demolitions man. All members should be cross-trained on all procedures. Demolitions are prepared for use before arriving at the point of breach. An engineer platoon uses squads in series through the minefield to clear a lane for a company team. The platoon must rehearse reduction procedures until execution is flawless, quick, and technically safe. During reduction, the engineer platoon will be exposed in the lane for 5 minutes or more depending on the mission, the minefield depth, and the platoon's level of training.

- **Buried minefield.** Manually reducing a buried minefield is extremely difficult to perform as part of a breaching operation. If mine burrows are not easily seen, mine detectors and probes must be used to locate mines. Mines are then destroyed by hand-emplaced charges. As an alternative, mines can be removed by using a grappling hook and, if necessary, a tripod (*Figure C-21*). Using a tripod provides vertical lift on a mine, making it easier to pull the mine out of the hole.

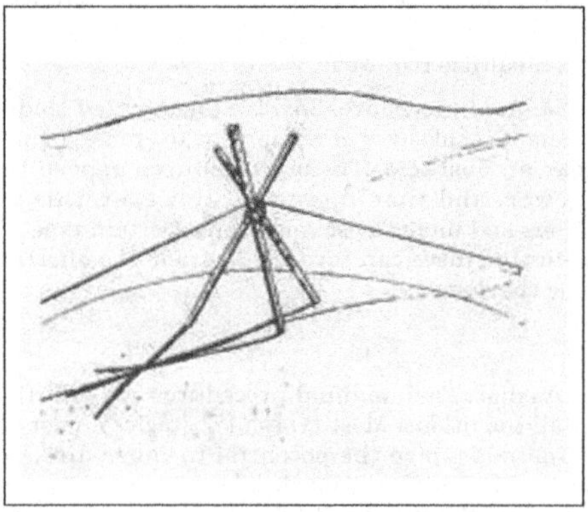

Figure C-21. Tripod

C-56. The platoon leader organizes soldiers into teams with distinct, rehearsed missions including grappling, detecting, marking, probing, and emplacing demolitions and detonating cord. The platoon is exposed in the obstacle for long periods of time.

Grappling Hook

C-57. The grappling hook (grapnel) is a multipurpose tool that is used for manual obstacle reduction. Soldiers use it to detonate mines from a standoff position by activating trip wires and AHDs. After the grapnel is used to clear trip wires in a lane, dismounted engineers can move through the minefield, visually locate surface-laid mines, and prepare mines for demolition. In buried minefields, soldiers grapple, and then enter the minefield with mine detectors and probes. There are two types of grapnel hooks—hand-thrown and weapon-launched.

- **Hand-thrown.** A 60+-meter light rope is attached to the grapnel for hand throwing. The throwing range is usually no more than 25 meters. The excess rope is used for the standoff distance when the thrower begins grappling. The thrower tosses the grapnel and seeks cover before the grapnel and rope touch the ground in case their impact detonates a mine. He then moves backward, reaches the end of the excess rope, takes cover, and begins grappling. Once the grapnel is recovered, the thrower moves forward to the original position, tosses the grapnel, and repeats the procedure at least twice. He then moves to the end of the grappled area and repeats this sequence through the depth of the minefield.
- **Weapon-launched.** A 150-meter lightweight rope is attached to a lightweight grapnel that is designed to be fired from an M16A1 or M16A2 rifle using an M855 cartridge. The grapnel is pushed onto the rifle muzzle with the opening of the retrieval-rope bag oriented toward the minefield. The firer is located 25 meters from the minefield's leading edge and aims the rifle muzzle at a 30- to 40-degree angle for maximum range. Once fired, the grapnel travels 75 to 100 meters from the firer's position. After the weapon-launched grapnel hook (WLGH) has been fired, the firer secures the rope, moves 60 meters from the minefield, moves into a prone position, and begins to grapple. The WLGH can be used only once to clear a minefield, but it can be reused up to 20 times for training because blanks are used to fire it.

C-58. Multiple grapplers can clear a lane of trip wires quickly and thoroughly, but they must time their efforts and follow procedures simultaneously. A hit on a trip wire or a pressure fuse can destroy the hook and the cord, so engineers should carry extras.

Appendix D

Breaching Scenario and Special Planning Considerations

This appendix provides a scenario showing how the breaching tenets may be applied when conducting a breaching operation in open terrain, such as a desert. This example is not an attempt to give a single, doctrinally correct solution but simply to provide a method that a unit may use to task-organize and approach this very difficult mission. Also discussed in this appendix are considerations that may be needed when conducting a breaching operation in restricted terrain (a defile, urban terrain, or the jungle). The following paragraphs are an example of a worst-case scenario in which a TF is given the mission to conduct a breaching operation against a well-equipped, well-prepared enemy defense.

SAMPLE BREACHING OPERATION

D-1. TF 1-12 is a mechanized infantry battalion that is task-organized with two tank companies from TF 3-71 (armored battalion). It just received a mission to secure OBJ SPARROW as part of a brigade attack to destroy a motorized infantry battalion (MIBN) (see *Figure D-1,* page D-2).

D-2. TF 1-12 received intelligence products, including the brigade SITEMP and a series of WOs, before issuing the actual OPORD. The TF staff, in particular the S2 and the engineer, revised and refined the brigade SITEMP based on their analyses of the terrain and the enemy (see *Figure D-2,* page D-3).

D-3. Upon receipt of the brigade OPORD, TF 1-12 learns that—

- The brigade has task-organized it with an engineer recon team to augment its recon capability and its habitually associated engineer company. An additional engineer line platoon has been attached to the engineer company, giving the TF a total of three engineer line platoons.
- The brigade has given TF 1-11 the mission to fix enemy forces north of OBJ SPARROW to prevent them from repositioning.
- The brigade plans to disrupt the enemy's counterattack force by using area-denial artillery munition (ADAM)/remote antiarmor mine (RAAM) SCATMINEs along the high-speed route to OBJ SPARROW. The counterattack force has been located by intelligence sources.
- The brigade plans to have the engineer battalion (-) follow TF 1-12's assault force to assume control of the lanes. The engineer battalion will create additional lanes, if necessary, and upgrade/maintain the lanes and marking pattern to facilitate the passage of TF 3-71.

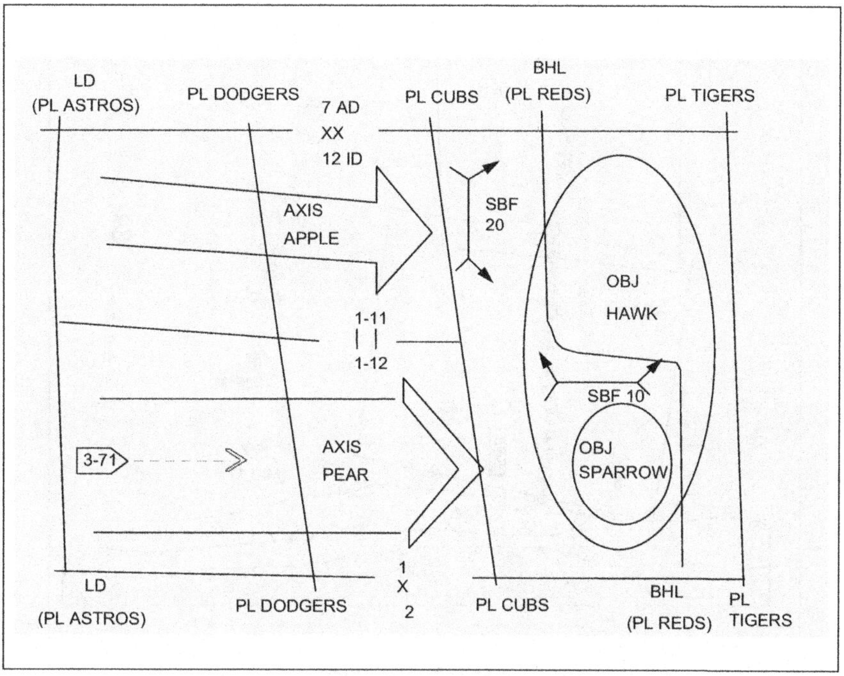

Figure D-1. Brigade Maneuver Graphics

- The brigade has tasked it to seize OBJ SPARROW and occupy SBF 10 with at least two platoons to fix enemy forces north of OBJ SPARROW.
- The brigade has tasked it to create two lanes in the obstacle system to support the passage of TF 3-71 through the created lanes.

D-4. Based on the commander's guidance, the TF staff developed three COAs. The third COA was chosen based on war gaming. It required a company team (assault force), consisting of one tank platoon (the platoon has two tank plows mounted) and two mechanized infantry platoons, to seize OBJ SPARROW. The staff decided that two lanes were necessary in the obstacle system, vicinity grid AA123456, based on the specified task given by the brigade. Because of the lane requirement and the obstacle type expected to be encountered, the staff decided that the breach force must consist of at least two tanks with MCBs, two tanks with MCRs, and two MICLICs. The breach force must also have two engineer platoons to create, proof, and mark the lanes and to provide redundancy of assets. The staff considered the threat to the breach force to be—

- The infantry platoon and AT weapons on the southern flank.

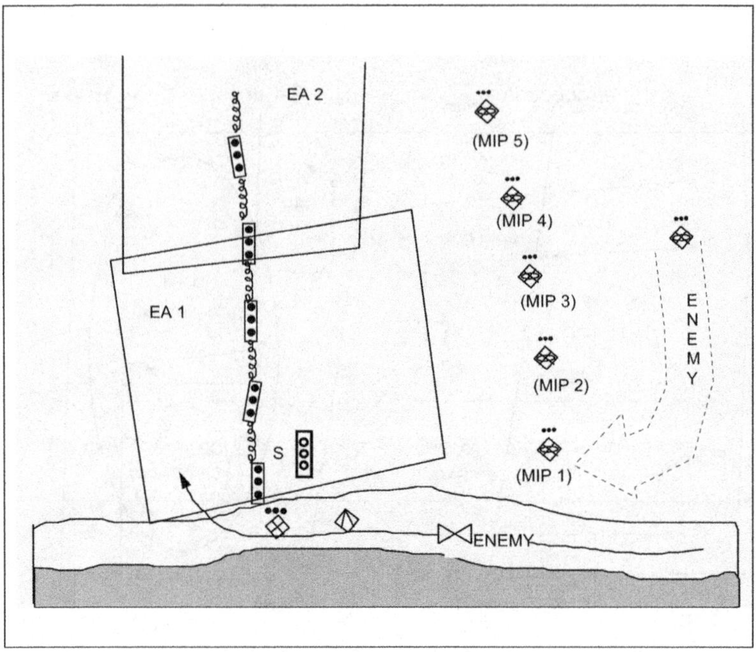

Figure D-2. TF SITEMP

- The southern motorized infantry platoon (MIP), which the support force may not be able to suppress because its forces are masking the breach force.
- The MIP-size counterattack force.

D-5. In this COA, the staff task-organized one tank platoon and one mechanized infantry platoon to the breach force (in addition to the MCBs and MCRs) to provide local security and redundant suppression. The staff determined from its terrain and enemy analyses that the support force's mission is to suppress the four southern MIPs to facilitate reduction operations. To accomplish this, the staff task-organizes two company teams to the support force. Each company team in the support force has one tank plow to reduce any obstacles it encounters en route to the SBFs (see *Figure D-3,* page D-4).

D-6. The staff further refines the COA by developing an initial plan to achieve the following breaching fundamentals, to include assigning tasks to the various subordinate forces (see *Figure D-4,* page D-5):

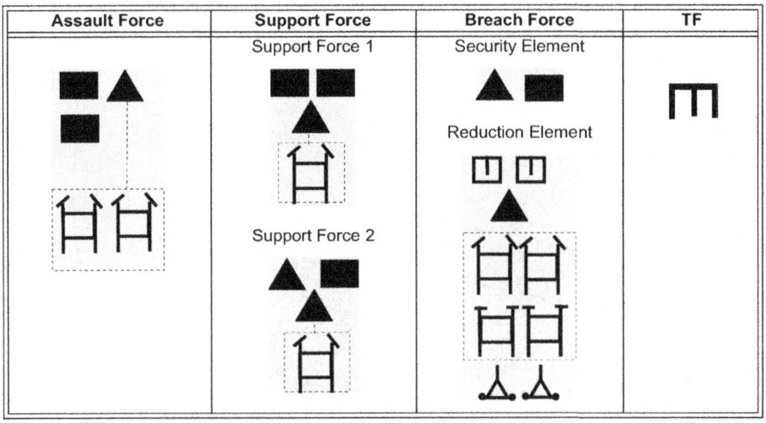

Figure D-3. Initial Breaching Organization

- **Suppress**.
  - Support force 1 will occupy SBF 1 and initially orient its fires on MIPs 1 and 2. When the breach force is committed, it will shift its fires to orient on MIPs 2 and 3. Fires will be lifted on the order of the assault-force commander.
  - Support force 2 will occupy SBF 2 and orient its fires on MIPs 3 and 4. Fires will be lifted on the order of the assault-force commander.
  - The breach-force security element will occupy SBF 3 and orient fires on MIP 1. The security element is responsible for local security at the point of breach.
  - The fire-support element (FSE) will ensure that CFZ 1 is activated as the breach force moves to the point of breach.
  - The support force will call for and adjust artillery targets AG7001 through AG7003 when the lead elements of the support force cross PL CUBS.
- **Obscure**.
  - The support force will call for and adjust smoke target AG1210 when the lead elements of the support force cross PL CUBS. The support force will adjust this target only if the smoke is ineffective.
  - The smoke platoon (task-organized to the TF for this mission) will employ its generators to the north of support force 2. Smoke operations begin when support force 2 crosses PL CUBS.
- **Secure**. The breach force (security element) will occupy SBF 4 and orient fires on the infantry platoon and AT weapons on the south wall.

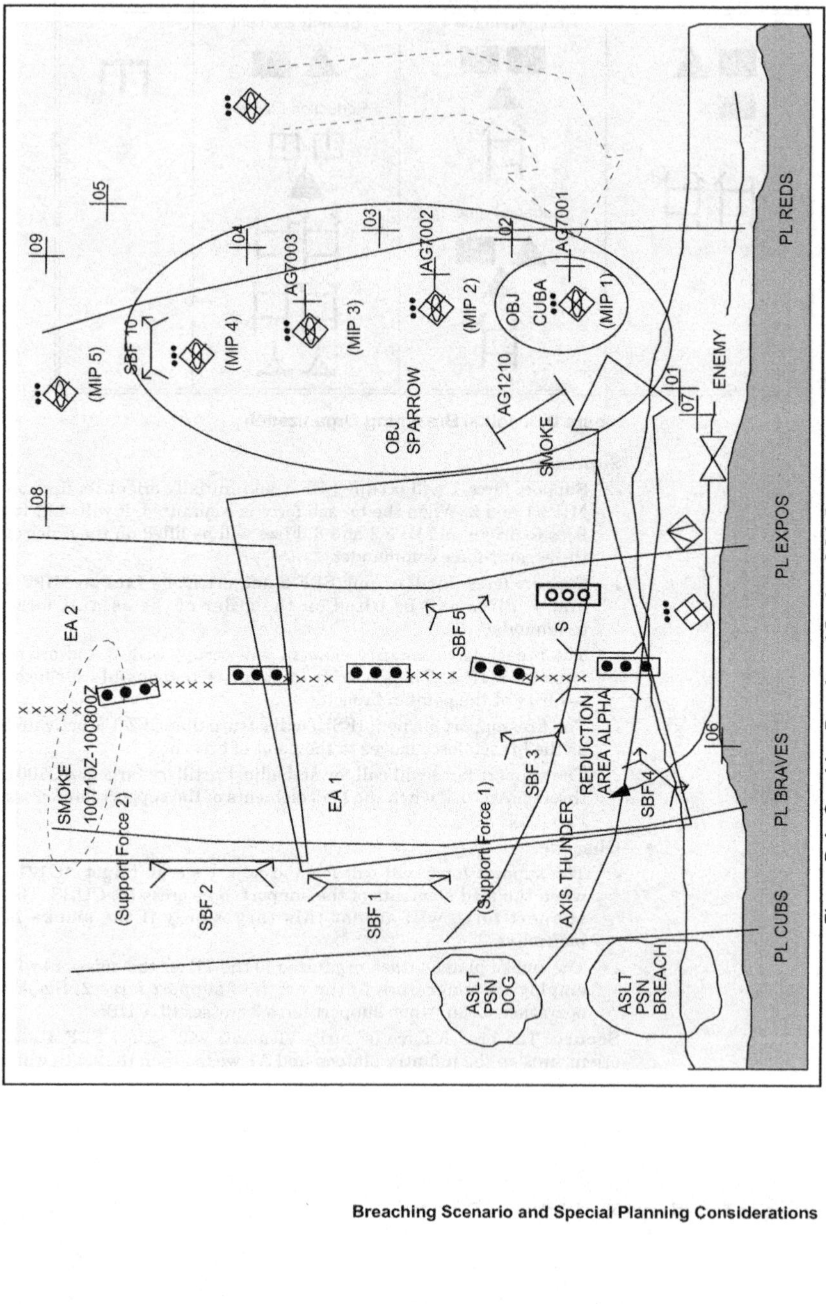

Figure D-4. Graphics Developed During COA Development (Overlaid on SITEMP)

- **Reduce.**
  - The breach-force reduction element will move from assault position (ASLT PSN) BREACH when the commitment criteria in reduction area ALPHA has been met. In this scenario, it will create, proof, and mark (initial lane-marking pattern) two lanes in the obstacle system, at least 800 meters apart, within reduction area ALPHA. This considers the enemy's use of artillery-delivered SCATMINEs to close the lanes.
  - The breach-force security element will occupy SBF 5 with one platoon orienting fires on MIPs 2 and 3 to provide enemy suppression when one lane is created in the tactical obstacles.
- **Assault.**
  - The assault force will attack along axis THUNDER to initially seize OBJ CUBA (point of penetration) and then the rest of OBJ SPARROW.
  - The assault force will occupy SBF 10 with at least two platoons oriented on MIP 5.

D-7. During COA analysis, modifications are made to the original COA. This includes task-organizing an engineer platoon to the assault force to help reduce protective obstacles and mark lanes for the assault and follow-on forces (see *Figure D-5*).

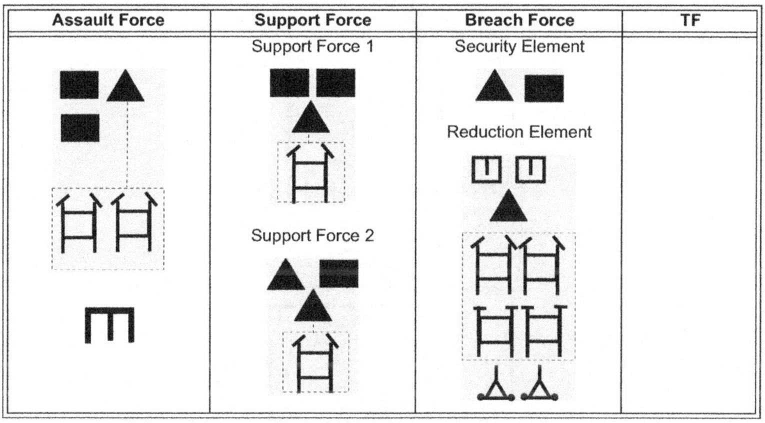

Figure D-5. Final Task Organization

D-8. Additionally, the staff develops a decision point for committing the breach force and a plan to collect the necessary information to support the decision point (see *Table D-1*). *Figure D-6* shows the location of each NAI.

Table D-1. Plan to Support the Decision Point

Decision Criteria	NAI	Responsibility
MIP in OBJ CUBA must be destroyed.	NAI 300	TF scouts
Enemy south of TRP 5 must be suppressed.	NAI 301	Support force commander
Point of breach must be effectively obscured.	NAI 302	Support force commander
Support force must have a minimum of 16 M1s/M2s available.		Support force commander
Breach force must have a minimum of one engineer platoon, 1 MICLIC/ MCB, and 6 M1s/M2s.		Breach force commander
CFZ 1 must be activated.		TOC (FSE)

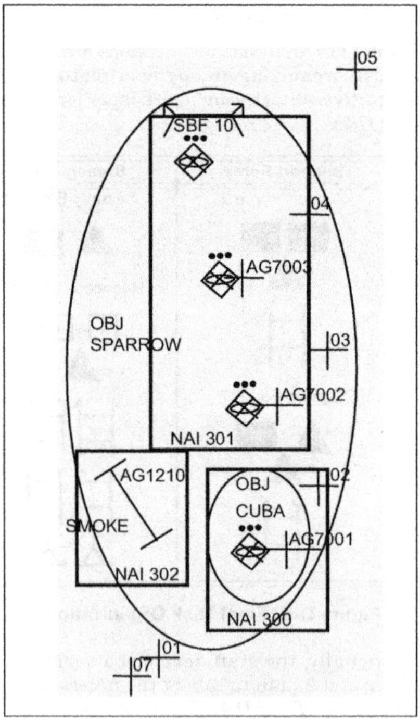

Figure D-6. NAIs to Support the Decision Point

D-9. The commander specifies the following C^2 structure of the mission based on the scheme of maneuver in the COA he approved:

- The assault force will be comprised of Team A/3-71 (mechanized heavy).
- The breach force will be comprised of Team B/3-71 (tank heavy) and A/ 55th Engineer Battalion. The Team B commander will be the breach-force and security-element commander. The engineer commander will be the reduction-element commander.
- The support force will be comprised of Team C/1-12 (mechanized heavy) and Team D/1-12 (tank heavy). The support-force commander will be the TF S3.

D-10. Preparation for the breaching operation involves numerous rehearsals, beginning at the lowest levels and culminating into a full-dress, combined-arms rehearsal. S2 reports that the TF received the confirmed location of the tactical obstacles in the vicinity of the point of breach. The report indicated that the minefields are comprised of three rows of buried SB-MV mines (Italian AT mines), with a uniform 4.5 meters spacing between mines and 70 meters between rows. One row of concertina wire is positioned 40 meters to the rear of each tactical minefield. Additionally, the enemy has emplaced protective minefields, which are comprised of AP mines only, 400 meters in front of each platoon position. The enemy has tanks and infantry fighting vehicles (IFVs) overwatching tactical and protective obstacles. Reports indicate that the enemy has not prepared dismounted positions.

NOTE: The assault-force commander decides to conduct a mounted assault based on information from the intelligence report and the terrain analysis he conducted. The terrain analysis indicates that tank plows will effectively reduce the enemy's protective obstacles.

D-11. This information is quickly relayed to subordinate units and all rehearsal sites reflect these obstacles. The commander pays particular attention to the status and location of reduction assets throughout preparations for the attack. This results in an MCB being removed from an inoperable M1 at the unit maintenance collection point (UMCP) to an operable M1 in the assault force.

D-12. The TF crosses the LD at 95 percent strength (all MCBs, MICLICs, and MCRs are operational). Before crossing PL DODGERS, the TF scouts report that three of the four vehicles in NAI 300 have been destroyed by Copperhead munitions. The commander reminds the S3 that destroying the remaining vehicles is his first priority.

D-13. As the support force crosses PL CUBS, the TF S3 reports that he has lost two M2s to indirect fires and one M1 to maintenance and is beginning to occupy SBFs 1 and 2. Artillery units begin suppressive fires and begin to build obscuration smoke. As the support force begins to suppress the enemy with direct and indirect fires, the breach and assault forces occupy ASLT PSNs BREACH and DOG, respectively, and conduct final preparations as necessary. The tactical operations center (TOC) informs the commander that CFZ 1 will be activated when the breach force is committed. The breach-force commander reports that all reduction assets are operable, and the TF scouts report that the final vehicle in OBJ CUBA has been destroyed. The TF S3 reports that obscuration and suppression have been achieved and that combat power is 6 M1s and 13 M2s.

D-14. Having met the commitment criteria, the commander orders the breach force to AA123456 to begin obstacle reduction. The breach-force commander initially moves his security element to SBFs 3 and 4. The reduction-element commander also moves forward with his platoon leaders to recon the obstacle and confirm/refine the plan to reduce the obstacle. Once the security element is set in SBFs 3 and 4, the reduction element moves forward to reduce the obstacle. When the first lane is created and proofed, the breach-force commander moves a portion of the security element from SBF 3 to SBF 5. The assault-force commander begins movement from the assault position to the far-recognition marker that has been reported by the reduction-element commander. Upon reaching the far-recognition marker, the assault-force commander can observe the final-approach marker and the marked lane.

D-15. The assault force moves through lanes in the tactical obstacles and sets its platoons in the SBFs. It immediately begins to create two lanes in the protective obstacles with the MCBs. When the lanes are complete, the assault-force commander maneuvers the two tank platoons to SBF 10, while the mechanized company team continues to suppress the remaining enemy vehicles. The engineer platoon remains at protective obstacles to mark lanes in preparation for passing follow-on forces.

D-16. The engineer battalion S3 is immediately behind the assault force. He takes control of the reduction element that has created, proofed, and marked two lanes in the tactical obstacles by this time. The S3 moves a portion of the reduction assets away from the lanes to counter any SCATMINEs that the enemy may use to close the lanes. The S3 prepares to assist with TF 3-71's passage through the lanes. When the assault force reports that it has created a safe route to PL REDS, the brigade commander orders TF 3-71 to pass through TF 1-12 and continue the attack.

CONSIDERATIONS FOR RESTRICTED TERRAIN

D-17. The breaching tenets discussed in *Chapter 1* apply to breaching operations in restricted terrain; however, the application of these tenets is different from a breaching operation conducted in open terrain. The following considerations should be made when breaching in restricted terrain:

- OBSTINTEL collection is critical when conducting breaching operations in restricted terrain. It is essential that the breach force knows what types of obstacles are in place and knows the exact composition of any complex obstacle. With this information, the breach-force commander can ensure that the reduction assets necessary to reduce the obstacle are available and placed in the right order in the movement formation. The breach force's ability to move reduction assets within the formation may be extremely limited due to the terrain's constricting nature.
- Recon assets tasked to obtain OBSTINTEL should be prepared to recon different types of obstacles than would normally be found in open terrain. For example, in urban or mountainous terrain, the

enemy may use rubble obstacles, such as the *rock drop*. Additionally, recon assets are much more likely to encounter field-expedient or booby-trapped obstacles when conducting operations in restricted terrain. Recon assets used to collect OBSTINTEL must be able to recommend reduction methods to be used by the breach force when reducing the obstacle.

D-18. The proper application of the following breaching fundamentals is essential to the success of breaching operations conducted in restricted terrain. As with other breaching tenets, breaching fundamentals are normally applied differently in restricted terrain than in open terrain.

- **Suppress**. The need to prevent or degrade enemy fires and observation of friendly forces remains unchanged when breaching in restricted terrain. However, unlike a breaching operation conducted in open terrain in which the support force can suppress the enemy with long-range fires, massed indirect fires, and smoke, suppression will normally be achieved through close-in fighting. Forces tasked with suppressing the enemy may find it necessary to seize terrain at or on the far side of the obstacle to achieve their mission. In all probability, the fighting to achieve enemy suppression will be decentralized. Therefore, the ability to use indirect fires will be restricted due to the close proximity of friendly and enemy forces.
- **Obscure**. When trying to obscure the enemy with indirect fires, the same problems exist as when trying to suppress the enemy. Using mortars (because of their ability to fire high-level trajectory), smoke pots, and smoke grenades may be more effective. The potential for degrading friendly-force operations when using obscurants is greater because of the close-in nature of the fighting that occurs in restricted terrain.
- **Secure**. The responsibility to secure the reduction area may normally be given to the support force when conducting breaching operations in restricted terrain. Because the support force may find it necessary to seize terrain on the far side of the obstacle to suppress the enemy, securing the reduction area may be accomplished at the same time. When this situation occurs, the amount of maneuver forces within the breach force may be less than what is normally necessary when breaching in open terrain.
- **Reduce**. The ability to create more than one lane and to move reduction assets around each other may be severely restricted when breaching in restricted terrain. For this reason, accurate OBSTINTEL and a well-thought-out, well-rehearsed reduction plan are critical.
- **Assault**. As discussed earlier, the number of lanes available to the assault force may be less than what is actually needed. For example, a battalion-size assault force may be restricted to using a single lane through an obstacle due to the restricted terrain. For this reason, the assault-force commander must ensure that the sequencing of forces through the lane is appropriate to achieve the mission.
- **Mass**. Although mass remains a critical tenet for breaching operations in restricted terrain, the ability to achieve mass may be difficult and time-consuming. The assault force will attempt to

achieve a 3:1 combat-power ratio. However, due to the restricted nature of the terrain and the limited number of lanes that may be created, the assault force may take longer to get that much combat power to the far side of the obstacle. Units must consider ways to minimize the risk to the assault force as it builds its combat power on the far side of the obstacle.

- **Synchronize**. The reverse-planning process described in *Chapter 1* remains the key to successful breaching operations in restricted terrain. However, effective C^2, particularly the controlling of personnel and equipment around the point of breach, is critical for successful operations. The terrain's restricted nature magnifies mistakes made in bringing the proper reduction assets forward to the obstacle or in sequencing forces through created lanes. These mistakes result in increased time for the enemy to react to the breaching operation through repositioning and other means. For this reason, a well-rehearsed force is critical. All subordinate units must understand their mission, the overall timing of the operation, and how their actions relate to missions of other subordinate units.

CONSIDERATIONS FOR MILITARY OPERATIONS IN URBANIZED TERRAIN

D-19. Military operations in urbanized terrain (MOUT) is defined as all military actions that are planned and conducted on a topographical complex and its adjacent natural terrain where man-made construction is the dominant feature. Urbanized terrain affects the tactical options available to the commander and requires a thorough knowledge of unique terrain characteristics, detailed planning down to the smallest unit level, and sound leadership at all levels. The complexities of the urban environment, such as line-of-sight (LOS) restrictions, inherent fortifications, limited intelligence, densely constructed areas, and the presence of noncombatants, restricts current military technology. US forces do not possess the overwhelming technology advantages in an urban environment as in other environments.

D-20. A three-dimensional view of the terrain is essential to gain a complete understanding of observation and AAs. The close fight (inside the building) is difficult and decentralized, with control of the movement from room to room most often in the hands of squad and platoon leaders. Success in the movement from building to building is a combined-arms fight, which requires the active involvement of a battalion command post. Seeing the terrain in a three-dimensional view will identify opportunities for employment of all the battalion's combat power and enable the battalion to set the conditions for successful platoon and company assaults.

PLANNING

D-21. Close-quarters clearing techniques are designed to be executed by standard four-man breach and clearing teams. Therefore, the size of the breach organization will be significantly smaller than those involved when fighting occurs in open terrain. The reduction element must work with the breach leader to formulate the assault plan and to ensure that all planning considerations are completed.

D-22. The reduction element is responsible for the following tasks during the planning phase:

- Accessing all intelligence data on potential breach points.
- Helping the command element select the best method of entry (mechanical, shotgun ballistic breach, or explosive). (See *Figures D-7 through D-9*, and *Figure D-10*, page D-14)
- Assembling and preparing tools and equipment and constructing explosive entry charges and diversionary charges if required.
- Briefing team members.
- Rehearsing.
- Conducting a back brief of team members.
- Conducting a precombat inspection (PCI).

D-23. The commander, breacher, and other essential personnel assemble and analyze all available information, data, and intelligence pertaining to the intended target. This involves examining the design construction and material makeup of the structure to determine the best primary and alternate points of breach.

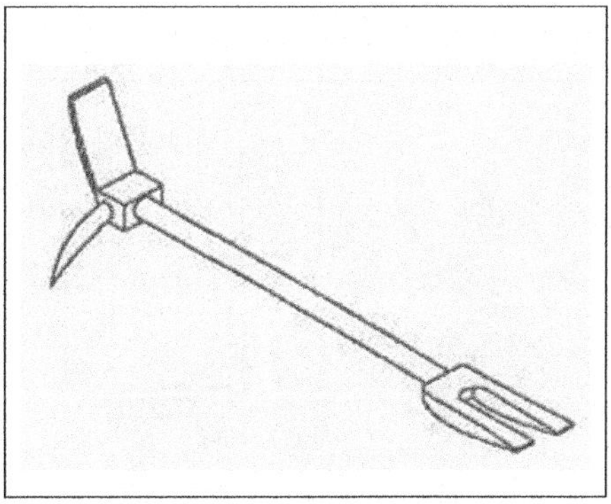

Figure D-7. Hooligan Tool

METHOD OF ENTRY

D-24. Once the point of breach has been identified, the reduction element leader continues to assemble and analyze all available information, data, and intelligence pertaining to the target structure. This involves a detailed examination of each point of breach's design, construction, and material makeup to determine the best primary, secondary, and tertiary breaching method. Again, tactics to be employed may dictate or rule out specific entry points.

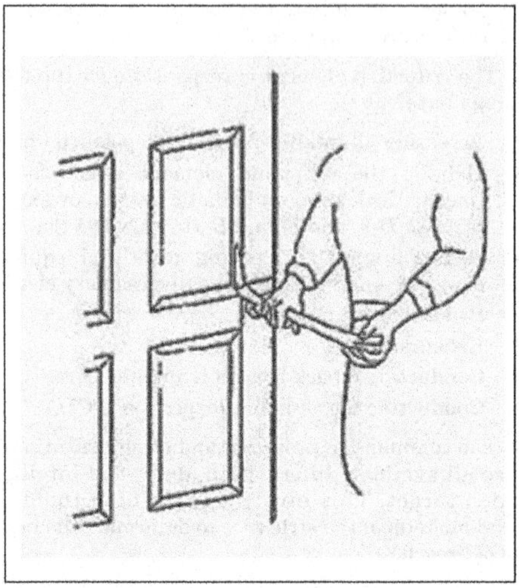

Figure D-8. Using a Hooligan Tool Against an Inward-Opening Door

Figure D-9. Attacking a Dead Bolt with a Sledgehammer

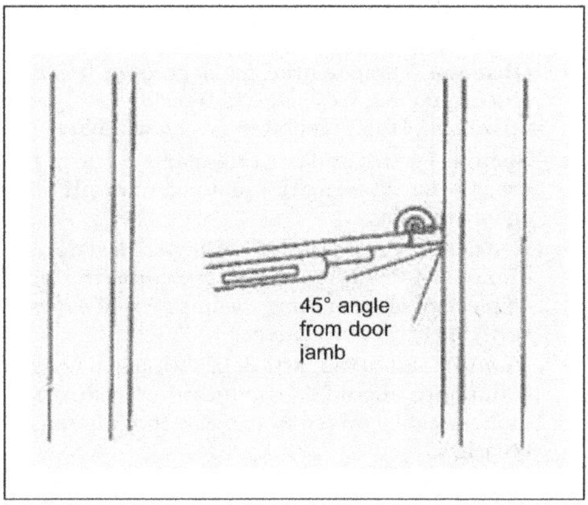

Figure D-10. Using a Shotgun Against a Door Lock

OBSTACLE INTELLIGENCE

D-25. Collecting the following OBSTINTEL is critical during MOUT:

- What type of mission is it?
- Are drawings or blueprints of the structure available?
- Are photographs of the point of breach and adjacent structures available?
- Can R&S personnel give a detailed description of the point of breach?
- Can counterintelligence personnel provide any information?
- Are there any human intelligence sources (agents, friendly forces, evacuees, prisoners, tourists, workers, businessmen, travelers) that can provide information relating to the point of breach?
- What type of shielding is available at potential breach points (natural or man-made)?
- How close will the assault force be positioned to the breach points?
- Do drawings of the target (if available) include gas, power lines, and other similar hazards?
- Is there a history of booby-trap use?
- Will the weather affect charge adhesion methods?

BREACHING FUNDAMENTALS

D-26. The proper application of the following breaching fundamentals is essential to ensure success in breaching conducted during MOUT. As with other breaching tenets, the breaching fundamentals are normally applied differently in urbanized terrain than in open terrain.

- **Suppression**. Maneuvering to achieve suppression of the enemy will be decentralized. Therefore, the ability to use indirect fires will be restricted due to the close proximity of friendly and enemy forces and civilians.

- **Obscure**. The potential for degrading friendly-force operations with obscurants is greater due to the close-in fighting that occurs in built-up terrain. Hand-emplaced smoke will likely be the most effective.
- **Secure**. Breach and clearing teams secure a foothold near the point of breach. Local security must consider all possible locations for the enemy to hide.
- **Reduce**. Redundant systems are critical during MOUT. The reduction element is responsible for preparing equipment/charges for emplacement, searching for improvised explosive devices (IEDs), and verifying target intelligence.
- **Assault**. The actual attack of the objective using close-quarters battle techniques should be conducted according to the assault plan. The unit should be prepared to conduct additional planned and unplanned breaches.

Appendix E

Route Clearance

Route clearance is a combined-arms operation typically executed by an infantry company or a cavalry troop. Units must clear lines of communication (LOC) of obstacles and enemy activity that disrupt battle-space circulation. Units must conduct route clearance to ensure that LOC enable safe passage of combat, combat-support (CS), and CSS organizations. Clearance operations are normally conducted in a low-threat environment.

METHODS AND LEVELS

E-1. The information gathered from the IPB and the recon effort determines the method and type of route clearance to conduct. The determination is based on the situation, the time available, the threat level, and available assets.

E-2. During stability and/or support actions, it is recommended that former-warring-faction (FWF), engineer-equivalent clearance teams precede US clearance teams within the FWF's AO. Do not assume that FWF clearance teams will be thorough in their clearance operation. Treat the route as unsafe until US or allied force clearance teams have proofed the route to confirm that it is cleared.

CLEARANCE METHODS

E-3. There are three methods of route clearance—linear, combat, and combat route. The method employed depends on the situation, the time available, and the clearance assets available. The maneuver force should always establish static security positions at critical locations following the completion of route clearance.

- **Linear**. In linear clearance (*Figure E-1,* page E-2), the clearance begins at checkpoint 1 and is completed at checkpoint 2. This method provides the best assurance of route clearance. Although this is an effective method, it is not the most secure method in a high-threat environment. It is also time-intensive and constrains the maneuver commander's flexibility.
- **Combat**. Whereas linear clearance focuses on a specific route, combat clearance (*Figure E-2,* page E-2) focuses on specific points along a route. As mentioned previously, the IPB can identify likely areas for obstacle and ambush locations. These areas become NAIs or objectives for combat clearance missions. The combat clearance method divides a route into sections according to the number of suspected high-threat areas. Once the unit clears these areas, the route is considered clear. Combat forces can patrol the route from these objectives to ensure

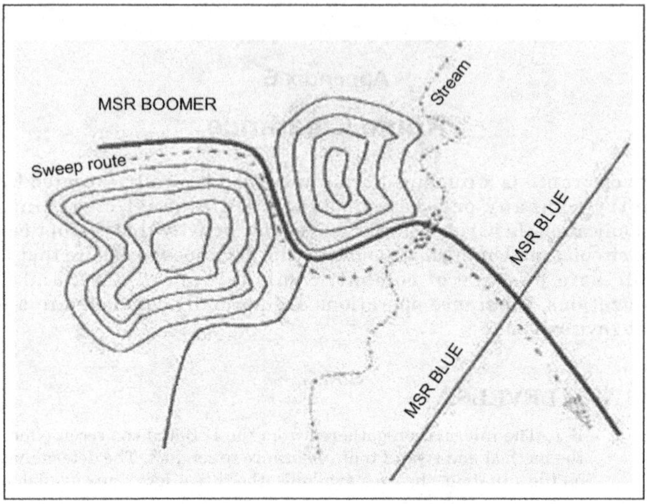

Figure E-1. Linear Clearance Method

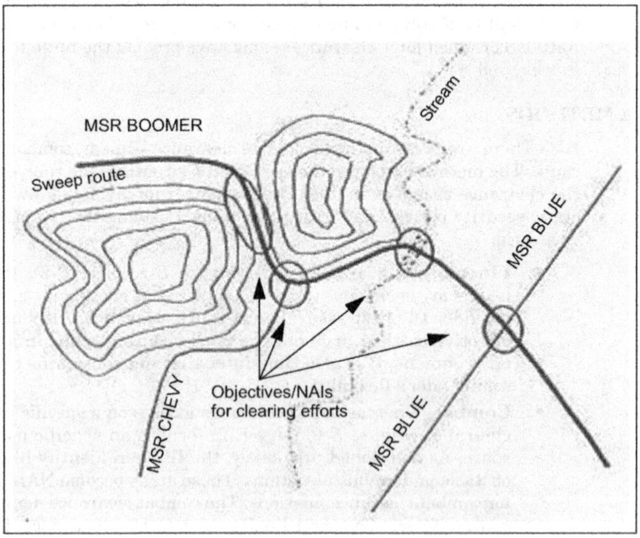

Figure E-2. Combat Clearance Method

that the route is secure and, if necessary, the sweep element can sweep the surrounding area if a minefield is found. Following the seizure of these objectives, the commander must assume a moderate risk that the S2 and the force engineer have identified all high-threat areas and that the route is clear of obstacles.

- **Combat route**. This method combines the clearance capabilities of the linear method with the security and surprise elements of the combat method. It is a two-phase, force-intensive operation and may require a battalion-size element, depending on the route's length. First, identified high-threat areas are targeted, secured, and cleared of obstacles and enemy forces before a sweep team moves. The sweep team then travels down the road and clears obstacles missed or not identified during the IPB/engineer battlefield assessment (EBA). The advantage of this method is that the TF commander immediately secures MSRs and then finds the enemy, confident that MSRs are relatively safe.

SWEEP LEVELS

E-4. The four levels of sweep operations are discussed below. All four levels can be modified to meet the unit's time and equipment limitations, but the commander assumes greater risk when the clearance level is modified. All four sweep levels can be used with any of the three clearance methods (linear, combat, and combat route).

- **Level one**. This is the fastest, riskiest type of route sweep. It is ideal for an armored/mechanized team. The sweep relies primarily on visual detection, using thermal or infrared sights or the naked eye for mine identification. It is followed immediately by a secondary, mechanical detection system such as MCRs. The sweep team is a squad-size or larger element that is task-organized with mine detectors, demolitions, and possibly an interim vehicle-mounted mine detector (IVMMD). The sweep team looks for minefields along the road width of a route. Security and support teams consist of a maneuver platoon to provide overwatching fire. The primary objective of a level one sweep is speed, with the sweep team moving at 5 to 8 miles per hour. The sweep team identifies immediate risks to traffic, neutralizes those risks, and continues on with the mission. If mine rollers are not available, a sandbagged 2 1/2-ton truck driven backwards can be used for a level one sweep, but only as a last resort. In this situation, the mine rollers are only a means of detection, not breaching. Mine rakes or plows are not always satisfactory substitutes because they most likely will destroy road surfaces.
- **Level two**. A level two sweep uses electronic measures as the primary detection method in high-threat areas. These areas include intersections, choke points, and areas within 10 meters of wood lines. A level two sweep employs more caution and forces the unit to update the IPB before beginning the mission. This level of operation employs a company team for security and C^2. Dismounted forces clear and secure the flanks and the far side of an identified minefield, while an engineer squad clears the road area.

- **Level three**. A level three sweep is more in-depth and more time-consuming. The sweep team must examine the route's entire width, including the shoulders and ditches. This technique ensures that follow-on forces are protected if they are forced to the side of the road. The security and support element also moves, mounted or dismounted, to provide rapid response and security. The sweep team provides a route recon report that updates current maps and further identifies high-threat areas. Staffs gain valuable information from this report for future operations.
- **Level four**. A level four sweep is very time-consuming. It relies on visual and electronic means for clearance. The platoon-size sweep team is dismounted to focus attention on the route's entire length. The company-size security element clears and secures the flanks and front of the sweep team (at least 100 meters in each direction in forested areas). This clears the area of off-route and command-detonated mines and allows the team to focus solely on the route. MCRs are used to proof the route after the sweep team passes through the area. Level four sweeps include route and area recons, and they are used when thoroughness and security, not speed, are critical to the mission.

EQUIPMENT

E-5. The IVMMD and the MinFlail are used to conduct route clearance.

INTERIM VEHICLE-MOUNTED MINE DETECTOR

E-6. The IVMMD (*Figure E-3*) is used in all levels of conflicts and stability and/or support actions. It is mounted on a blast- and fragmentation-survivable vehicle and is designed to detect and mark buried and surface-laid, metallic AT mines. The primary mission of the IVMMD is to detect mines during route-clearance operations. The system should not be operated in an environment where the enemy employs mines that are not pressure-fused.

System Components

E-7. A complete IVMMD consists of one mine-detection vehicle (MDV), one towing/mine-detection vehicle (T/MDV), three mine-detonation trailers, a spare-wheel module for the MDV, a spare-wheel module for the T/MDV, and a container of spare parts.

E-8. The MDV's only mission is to detect mines. It can negotiate vertical slopes up to a 20 percent grade. The MDV employs a 4-cylinder engine and a manual transmission to propel the 4.8-ton vehicle with a 3-meter-wide detection array. The detection array consists of two separate induction coils (one for the left side and one for the right side) that detect magnetic fields below the vehicle. The detection array is suspended between the two axles of the vehicle. When the detector encounters a metallic object, the operator is notified by an audible signal in the earphone and a visual signal on the dashboard that denotes which side array detected the object. The operator then stops the vehicle, backs it up, and reencounters the metallic object. (The MDV has two detection modes—the locate mode is used to identify the metal object, and the pinpoint mode is used to find the center of the object.) When

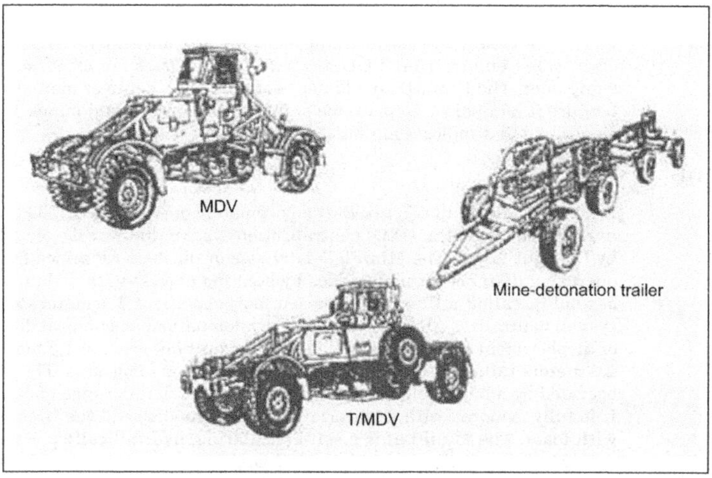

Figure E-3. IVMMD Components

the operator encounters the strongest signal, he activates the marking system (a nozzle mounted on the rear frame and centered on each detection array) that deploys a water-based ink onto the roadway.

E-9. The MDV is designed to produce very little ground pressure; however, it may detonate pressure-fused mines, depending on their fuse sensitivity. The MDV produces 27.9 pounds per square inch (psi) of ground pressure when the tires are inflated to 14.5 psi and 21.8 psi of ground pressure when the tires are inflated to 8.7 psi.

E-10. The T/MDV has a 6-cylinder engine and the same detection and marking system as the MDV. The T/MDV produces 49.8 psi of ground pressure when the tires are inflated to 29 psi and 23.4 psi of ground pressure when the tires are inflated to 8.7 psi. The T/MDV tows three mine-detonation trailers, and the recommended maximum operating speed while towing the trailers is 20 kph. The T/MDV (with trailers) can negotiate vertical slopes up to a 20 percent grade; however, going down such slopes is difficult. The T/MDV must be in first gear, and the trailer brakes must be deployed to decrease the speed of the vehicle when going down a slope.

E-11. The mine-detonation trailers are very heavy and specifically designed to apply heavy ground pressure that initiates pressure-activated mines. Each trailer has two axles of different lengths so that the three trailers provide a full 3-meter-wide proofing capability behind the T/MDV. If a mine detonates underneath the trailers, the wheel bolts are designed to sheer so that repair is limited to replacing a single wheel.

Employment Concept

E-12. The IVMMD is used to support route-clearance operations. Clearance operations ensure that LOC are safe for the passage of personnel and equipment. The IVMMD should not be used during hours of limited visibility, because it hampers the operator's ability to see surface-laid mines and visual signatures that indicate mining activities.

MINIFLAIL

E-13. The MiniFlail (*Figure E-4*) is a remotely operated, LOS, AP-mine and unexploded-ordnance (UXO) neutralization system that was developed for use by US light forces. The MiniFlail detonates or disables AP mines from a safe operating distance. It neutralizes by striking objects with a rotating chain assembly, called a flail, and clears a footpath about 1.1 meters wide. The system neutralizes AP mines and UXO by detonation, mechanical destruction, or displacement from the cleared lane. The MiniFlail is about 1.3 meters wide, 1.3 meters tall, and 3 meters long; it weighs 1,100 kilograms. The system is operated by a hand-held controller that has a maximum range of 300 meters. It is fully armored with a material similar to Kevlar, and the tires are filled with foam. The MiniFlail is a self-articulating, hydraulically powered shaft with 84 chains; each chain is 0.5 meter long.

Figure E-4. MiniFlail

PLANNING

E-14. The principles of breaching operations apply to developing and executing the route-clearance mission. The breaching tenets (intelligence, fundamentals, organization, mass, and synchronization) are the basis for planning.

INTELLIGENCE

E-15. Incorporating the IPB and METT-TC factors into route-clearance operations enables units to predict what the enemy will do and where it will do it. After the S2 and the engineer identify the most probable threat sites, the S2 designates them as NAIs. These NAIs are the focus of the recon effort.

Recon assets confirm the presence or absence of ambushes, UXO, and obstacles. The information gathered from the IPB and the recon effort determines the method and type of route clearance necessary. It also helps the commander determine any outside resources (explosive ordnance disposal [EOD] and special-operations forces [SOF]) that he may need.

FUNDAMENTALS

E-16. SOSRA may not be executed, but it is planned as it is in breaching operations. Units must be prepared to execute SOSRA fundamentals as necessary.

ORGANIZATION

E-17. Route-clearance operations are similar to breaching operations; therefore, commanders should plan and coordinate breaching fundamentals to suppress, obscure, secure, and reduce obstacles. Task-organizing for route clearance is also similar to that required for breaching operations. The assault force becomes the security element, the breach force becomes the sweep element, and the support force remains the same.

E-18. The significant difference between breaching and clearing operations is that breaching usually occurs during an attack (while under enemy fire) to project combat power to the far side of an obstacle. Route clearance focuses on opening LOC to ensure the safe passage of combat and support organizations within an AO. *Table E-1*, page E-8, shows a sample task organization for a route clearance.

MASS

E-19. Sufficient maneuver and engineer assets must be allocated to the sweep force. The length and width of the route and the type of clearance to be conducted determine the composition of the breach force. Clearing a Class A military road with the deliberate sweep technique requires at least two IVMMDs and/or two engineer squads due to the total lane width to be cleared and the requirement for rotating mine-detector operators. Depending on the type of sweep operation, the commander should plan for a 50 percent loss of sweep assets.

SYNCHRONIZATION

E-20. All aspects of synchronization should be implemented when planning route clearance. It is especially important that the clearance force conduct rehearsals at the combined-arms level. Rehearsals should include—

- Reaction to enemy contact, including ambushes.
- Communications exercise.
- Fire support (obscuration smoke, immediate suppression fires, CFZs, and no-fire area around the clearance site).
- CSS (casualty evacuation, marking materials, and demolitions resupply).

Table E-1. Sample Task Organization for a Route Clearance

Team	Support Force	Security Force	Sweep Force
Heavy	Mechanized infantry platoon with dismount capability Armor platoon FIST/COLT/Striker ADA section	Mechanized infantry platoon Engineer squad Mortar section Medical team (two ambulances) PSYOP team FIST MP element	Engineer platoon with organic vehicles Armor platoon with plows and rollers
Light/Heavy	Two infantry platoons (light) FIST/COLT/Striker ADA section	Bradley platoon with dismount capability Engineer squad 60-mm mortar section Medical team (two ambulances) PSYOP team FIST MP element	Engineer platoon with organic vehicles Armor platoon with plows and rollers
Light	Two infantry platoons (light) FO/COLT/Striker ADA section	AT/MP section with M60/MK19 mix 60-mm mortar section Medical team (two ambulances) PSYOP team FO MP element	Engineer squad (+) Infantry platoon (light) AT/MP section with M60/ MK19 mix

PLANNING CONSIDERATIONS

E-21. The purpose of breaching is to project combat power to the far side of an obstacle, and breaching usually occurs under enemy fire. Route-clearance operations focus on opening and maintaining LOC to ensure the safe passage of combat, CS, and CSS organizations. Like breaching, route-clearance operations require extensive Battlefield Operating System (BOS) coordination. The planning considerations shown in *Table E-2* should be used by brigade and battalion TFs when planning route-clearance operations. The obstacle template produced from this process determines which method and type of route clearance is required.

Table E-2. Planning Considerations for Route Clearance

INTELLIGENCE
Identify choke points, bridges, tunnels, critical road junctions, and built-up areas. These are the most likely locations for obstacle emplacement. However, depending on the enemy's overall mission, it may not always emplace obstacles at these locations. This is especially true if the enemy's goal is to psychologically disrupt convoys.
Maintain a situation map that reflects the most current intelligence information.
Maintain an incident map with a graphics overlay to facilitate a pattern analysis.
Maintain a threat order-of-battle database, such as how the enemy has disrupted LOC in the past.
Develop a detailed ISR plan that incorporates modern battle-space techniques and systems, such as ground sensors, forward-looking airborne radar, and satellite images. As a minimum— • Coordinate for UAV support, if available. • Develop infiltration routes to support recon and security at likely enemy ambush sites. • Develop an estimate of the impact to COBs (local nationals, NGOs, and PVOs). • Conduct a daily flight over the area to provide up-to-the-minute intelligence. • Coordinate with the USAF to periodically check the route (for example, use an AC-130 Specter gunship).
Establish liaison with the host nation, NGOs, and SOF.
Provide detailed OBSTINTEL, including— • Description of mines or explosive devices most likely encountered. • Composition and pattern of obstacles. • Enemy actions or techniques used during obstacle emplacement.
Provide intelligence updates on mine-hazard areas to company team leaders before departure. This should be in the form of a 1:50,000 enemy SITEMP overlay (confirmed and suspected/templated).
MANEUVER
Provide personnel for TCPs.
Secure flanks (at least 100 meters) and the far side of suspected and known obstacle locations.
Close the route to US-controlled traffic during route-clearance operations to minimize the target presented to enemy forces.
Identify and clear potential sniper positions before beginning obstacle reduction or clearance.
Provide security for the cleared route.
Plan the building of static security points along the cleared route to reduce the probability of reseeding.
FIRE SUPPORT
Plan smoke for templated obstacle locations.
Position mortars to ensure continuous coverage of the operation (one technique is to move under the control of the support force).
Prepare fires within the tactical rules of engagement.
Designate obstacle clearance sites as CFZs and no-fire areas.
Plan suppression fires on enemy elements capable of placing direct or indirect fires on the points of breach.

Table E-2. Planning Considerations for Route Clearance (continued)

MOBILITY/SURVIVABILITY
Report, clear, and mark obstacles and explosive devices to facilitate unimpeded movement.
Ensure that lane marking meets the standards directed by the supported commander and that materials and patterns are standard throughout the route.
Consider including road-repair equipment and material as part of the breach force (for example, a 5-ton dump truck filled with soil and an ACE to spread the soil).
Keep all radios, electronic equipment, and aviation assets at a safe distance during reduction operations.
Block uncleared roads and trails that branch from the route being cleared. This protects units from inadvertently traveling an uncleared route.
Debrief the chain of command and the TF S2 on the location, composition, and orientation of all obstacles cleared and encountered. This assists the S2 in IPB pattern analysis.
Provide detailed OBSTINTEL on minefields, including— • Description of mines or explosive devices most likely to be encountered. • Composition and pattern of obstacles. • Enemy actions or techniques used during obstacle emplacement.
ADA
Assist the S2 with the threat-air portion of the IPB.
Use the following passive air-defense measures: • Eliminate glare by using mud, tape, cardboard, or camouflage nets to cover headlights, mirrors, and portions of windshields. • Reduce dust clouds by reducing speed. • Plan routes that offer natural concealment. • Use air guards.
Increase the distance between vehicles.
Incorporate Stinger missile teams into the support force.
CSS
Ensure that clearance operations are supported by a logistical/CHS package.
Plan for air and ground evacuation of casualties. The preferred evacuation method is by air; the routine method is by ground.
Conduct an air-mission brief with air-ambulance assets, including pickup zones and markers.
Rehearse procedures for evacuation requests.
Ensure that the medical team consists of one or two ambulances.
Locate the medical team with the support force.
Identify the ambulance exchange point along the route to be cleared.
Ensure that appropriate personnel wear flak vests or IBASIC (*Figure E-5*, page E-12).
Ensure that all vehicles have tow cables in the front and the rear, and plan recovery assets for extraction purposes.
Ensure that all vehicles carrying troops have hardening (sandbags on floors and sides).
Provide MP and explosive-sniffing dogs to help in clearance operations and provide security for convoys during and after clearing operations.

Table E-2. Planning Considerations for Route Clearance (continued)

C^2
Designate a controlling, coordinating, and supporting HQ for route movement.
Designate, recognize, and include the following minefield indicators as part of company-team rehearsals: • Dead animals. • Avoidance by the local population. • Signs of digging. • Signs of concrete removal. • Disturbances in a road, such as holes or grooves. • Boxes or parcels placed along the road/shoulder. • Parked vehicles or bicycles without operators. • Wire on the road surface or extending onto the shoulders. • Evidence of vegetation disturbance along shoulders. • Evidence of mine-peculiar supplies such as wrenches, shipping plugs, wrapping paper or safety collars from fuses. • Posted signs that covertly alert the local population to the presence of mines. • Disturbances in tire tracks.
Designate a reserve force (at least platoon-size) that is mechanized or air-assault capable.
Ensure that proper rehearsals are planned and conducted. As a minimum, the unit should rehearse actions at the obstacle, actions on enemy contact, casualty evacuation, and control of COBs.
Ensure that the unit has a clear understanding of the mission, the intent, and the end state. For example, the clearing-unit commander should understand that his unit must clear the road width, including the shoulders, and secure the route.
Ensure that the unit— • Controls the movement of all personnel and equipment along the route. • Clears personnel within the minimum safe distance when clearing mines. • Prepares a risk assessment of the mission before issuing the OPORD. • Tracks the status of routes based on the amount of time since the route was cleared and the intelligence and enemy situations. • Tracks the progress of the clearance operation and integrates it into the maneuver and CSS plans. • Determines the route length, using clearly definable start and end points. • Coordinates with adjacent units, the host nation, NGOs, PVOs, and SOF. • Sets priorities for the route-clearance element.
SPECIAL OPERATIONS
Ensure that PSYOP/CA support the counterintelligence effort by conducting civilian interviews.
Direct civilians along the MSR to the displaced-personnel holding areas or along the routes that the unit has indicated for use.
Employ PSYOP/CA teams forward to disperse civilians and provide traffic management to isolate the route during clearance operations.

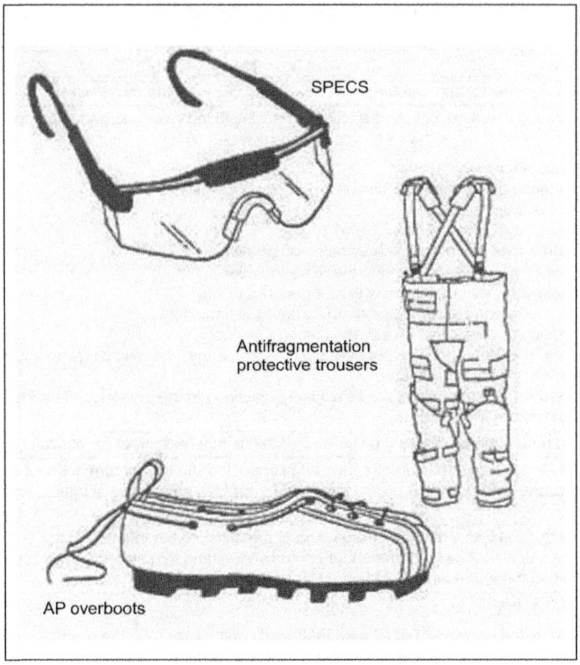

Figure E-5. IBASIC

TASK ORGANIZATION

E-22. A brigade or battalion TF normally conducts clearance-in-zone operations. To clear a route, the battalion TF focuses a company team as the main effort on the proposed MSR. *Table E-1*, page E-8, shows a sample task organization for a route clearance.

SUPPORT FORCE

E-23. The support force provides security for the breach force and neutralizes hostile forces. In rugged terrain or highly mined areas, moving the support force on the flanks may be too risky. In this situation, aviation assets may be used to provide flank security while ground forces provide rear security. The support force may also be tasked to search for suspected off-route mines.

SECURITY FORCE

E-24. The security force is prepared to maneuver and destroy enemy forces that the support force cannot neutralize. The security force may be comprised of maneuver and engineer forces augmented with a medical team, a psychological operations (PSYOP) team, and an EOD team. Additionally, the security force may be supported by a mortar section.

SWEEP FORCE

E-25. The sweep force is comprised of maneuver and engineer forces with detection equipment. The sweep force sweeps the route and reduces mine and explosive threats. As in breaching operations, the sweep force consists of a security element and a reduction element. The security element provides local security for the sweep force. The reduction element consists of sweep teams (personnel and equipment) to reduce obstacles and clear mines.

E-26. A sweep team is a trained detection team that searches for mines and explosive devices. The organization of the sweep team depends on the type of sweep mission, the length and width of the road to be cleared, the time and equipment available, and the surface composition (pavement, gravel, dirt) of the road to be swept. A dismounted platoon-size element can normally clear a 4.5-meter-wide path, and a squad-size element can normally clear a 1.5-meter-wide path. Mounted detection equipment, such as the IVMMD, can detect mines along a 3-meter-wide path. See *FM 20-32* for detailed information on the composition of a sweep team.

Glossary

AA avenue of approach

ACE armored combat earthmover

AD armored division

ADA air-defense artillery

ADAM area-denial artillery munition

AHD antihandling device

ammo ammunition

AO area of operations

AP antipersonnel

APC armored personnel carrier

APOBS Antipersonnel Obstacle Breaching System

ASLT assault

AT antitank

atk attack

ATTN attention

AVLB armored vehicle-launched bridge

AVLM armored vehicle-launched MICLIC

BFV Bradley fighting vehicle

BHL	battle handover line
BOS	battlefield operating system
C^2	command and control
CA	civil affairs
CAS	close air support
CCIR	commander's critical-information requirements
cdr	commander
CFZ	critical friendly zone
CHS	combat health support
CLAMS	Cleared-Lane Marking System
cm	centimeter(s)
cmd	command
COA	course of action
COB	civilians in the battle space
COLT	combat observation lasing team
CP	checkpoint
CS	combat support
CSS	combat service support
Czech	Czechoslovakia
DA	Department of the Army
det	detonated

dia	diameter
DPICM	dual-purpose improved conventional munition
DS	direct support
DST	decision-support template
ea	each
EA	engagement area
EBA	engineer battlefield assessment
EOD	explosive ordnance disposal
EOM	end of mission
FD	firing device
FECS	Field-Expedient Countermine System
FFIR	friendly forces information requirements
FIST	fire-support team
FM	field manual
FM	frequency-modulated
FO	foward observer
FRAGO	fragmentary order
FSE	fire-support element
FSO	fire-support officer
FWF	former warring faction
GPS	Global Positioning System

HE high explosive

HEMMS Hand-Emplaced Minefield Marking System

HEMTT heavy expanded mobility tactical truck

how howitzer

HQ headquarters

IBASIC Improved Body Armor System, Individual Countermine

ICP intelligence-collection plan

ID infantry division

IDA improved dog-bone assembly

IED improvised explosive device

IFV infantry fighting vehicle

IP improved plow

IPB intelligence preparation of the battlefield

IR information requirements

ISR intelligence, surveillance, and reconnaissance

IVMMD interim vehicle-mounted mine detector

km kilometer(s)

kph kilometers per hour

lb pound(s)

LD line of departure

ldr leader

LO liaison officer

LOC lines of communication

LOS line of sight

m meter(s)

MCAP mine-clearing/armor-protection

MCB mine-clearing blade

MCR mine-clearing roller

MCRP Marine Corps reference publication

MDMP military decision-making process

MDV mine detection vehicle

METT-TC mission, enemy, terrain, troops, time available, and civilian considerations

MIBN motorized infantry battalion

MIC motorized infantry company

MICLIC mine-clearing line charge

MIP motorized infantry platoon

MLC military load class

mm millimeter(s)

MOUT military operations in urbanized terrain

MP military police

MSR main supply route

NA not applicable

NAI named area of interest

NATO North Atlantic Treaty Organization

NBC nuclear, biological, chemical

NGO nongovernment organization

NSN national stock number

OBJ objective

OBSTINTEL obstacle intelligence

OP observation post

OPFOR opposing force

OPORD operation order

oz ounce(s)

PCI precombat inspection

PIR priority intelligence requirements

PL phase line

psi pounds per square inch

PSN position

PSYOP psychological operations

PVO private volunteer organization

R&S reconnaissance and surveillance

RAAM remote antiarmor mine

RCU remote control unit

RL	release line
RP	reference point
S2	Intelligence Officer (US Army)
S3	Operations and Training Officer (US Army)
SBF	support by fire
SCATMINE	scatterable mine
SITEMP	situation template
SOF	special-operations forces
SOP	standing operating procedures
SOSRA	suppress, obscure, secure, reduce, and assault
SP	start point
SPECS	Special Protective Eyewear Cylindrical System
SPOTREP	spot report
STANAG	Standardization Agreement
T/MDV	towing/mine-detection vehicle
TAI	targeted area of interest
TC	tank commander
TCP	traffic-control post
TF	task force
TM	technical manual
TOC	tactical operations center

TRADOC	United States Army Training and Doctrine Command
TRP	target reference point
TTP	tactics, techniques, and procedures
UAV	unmanned aerial vehicle
UK	United Kingdom
UMCP	unit maintenance collection point
US	United States
USAF	United States Air Force
UXO	unexploded ordnance
WLGH	weapon-launched grapnel hook
WO	warning order
wt	weight
Yug	Yugoslavia

Bibliography

DA Form 2028. *Recommended Changes to Publications and Blank Forms.* 1 February 1974.

FM 3-50. *Smoke Operations.* 4 December 1990.

FM 5-71-2. *Armored Task-Force Engineer Combat Operations.* 28 June 1996.

FM 5-71-3. *Brigade Engineer Combat Operations (Armored).* 3 October 1995.

FM 5-71-100. *Division Engineer Combat Operations.* 22 April 1993.

FM 5-105. *Topographic Operations.* To be published in the next six months.

FM 5-170. *Engineer Reconnaissance.* 5 May 1998.

FM 6-71. *Tactics, Techniques, and Procedures for Fire Support for the Combined Arms Commander.* 29 September 1994.

FM 6-121. *Tactics, Techniques, and Procedures for Field Artillery Target Acquisition.* 25 September 1990.

FM 7-10. *The Infantry Rifle Company.* 14 December 1990.

FM 17-98. *Scout Platoon.* 9 September 1994.

FM 20-32. *Mine/Countermine Operations.* 29 May 1998.

FM 34-130. *Intelligence Preparation of the Battlefield.* 8 July 1994.

FM 71-1. *Tank and Mechanized Infantry Company Team.* 26 January 1998.

FM 71-2. *The Tank and Mechanized Infantry Battalion Task Force.* 27 September 1988.

FM 71-3. *The Armored and Mechanized Infantry Brigade.* 8 January 1996.

FM 90-7. *Combined Arms Obstacle Integration.* 29 September 1994.

FM 100-5. *Operations.* 14 June 1993.

FM 100-7. *Decisive Force: The Army in Theater Operations.* 31 May 1995.

FM 101-5. *Staff Organization and Operations.* 31 May 1997.

FM 101-5-1. *Operational Terms and Graphics (MCRP 5-2A).* 30 September 1997.

STANAG 2036. *Land Mine Laying, Marking, Recording and Reporting Procedures.* Edition 5. 28 January 1999.

TM 5-6665-298-10. *Operator's Manual for AN/PSS-12 Mine Detecting Set (TM 5-6665-298-10).* 28 February 1995.

TM 9-1375-215-14&P. *Operator's, Unit, Direct Support and General Support Maintenance Manual (Including Repair Parts and Special Tools List) for Demolition Kit, Mine Clearing Line Charge (MICLIC)*. 31 January 1992.

Index

Field Manual
No. 5-10

Headquarters
Department of the Army
Washington, DC, 3 October 1995

Combat Engineer Platoon

TABLE OF CONTENTS

Page

PREFACE

Field Manual (FM) 5-10 is designed as a companion manual for *FM 5-34* and is structured using the following Battlefield Operating Systems (BOS) as a framework:

- Intelligence.
- Maneuver.
- Mobility and survivability.
- Fire support (FS).
- Air defense (AD).
- Combat service support (CSS).
- Command and control (C^2).

Each BOS is not discussed in detail. Instead, they are used to outline considerations for how the engineer platoon supports the total combined-arms (CA) effort. The bulk of the platoon's missions concentrates on mobility, countermobility, and survivability operations in support of the mobility and survivability BOS. However, the platoon leader and sergeant must have an understanding of the role their platoon plays in supporting all the BOS. More importantly, the platoon leader and sergeant must understand the impact of all the BOS on platoon planning, preparation, and execution.

This FM also has three appendices: *Appendix A* covers basic formations, movement techniques, and hand-and-arm signals; *Appendix B* contains checklists for security, precombat, and packing; and *Appendix C* covers case studies for mounted and dismounted breaching.

The proponent of this publication is Headquarters (HQ), United States (US) Army Engineer School. Send comments and recommendations on *Department of the Army (DA) Form 2028* directly to Commander, US Army Engineer School, ATTN: ATSE-T-PD-PM, Fort Leonard Wood, Missouri 65473-6650.

Unless this publication states otherwise, masculine nouns and pronouns do not refer exclusively to men.

CHAPTER 1
Intelligence

Intelligence is the key for preparing for the next battle. The analysis of intelligence, in detail, determines the impact of the enemy, weather, and terrain on operations. It is a continuous process that supports the planning for and execution of all operations. It is imperative that the platoon leader understand the intelligence process and his part in that process.

THREAT

The threat facing the US Army has gone through a quantum change with the breakup of the Warsaw Pact and the former Soviet Union. However, the majority of the forces we will face in mid- to high-intensity conflicts will use the "Soviet" model for their operations. *Figures 1-1 and 1-2, pages 1-2 and 1-3,* provide templates for the Soviet-style platoon and company defense as well as movement formations. A detailed analysis of threat mine-warfare capabilities and tactics can be found in *FM 90-13-1, Appendix A, and FM 20-32, Appendix C.*

RECONNAISSANCE

Reconnaissance is critical to the engineer. During offensive operations, the engineer must see the battlefield to plan breaching operations. History has proven that when a unit conducts CA rehearsals and obstacle reconnaissance it has greater success. The platoon leader should organize his platoon based on the reconnaissance mission the commander gives him. This frequently means providing expert engineers to assist the task force (TF) scouts. The platoon leader may also organize all or part of his platoon for a reconnaissance mission if there is a critical requirement for engineer-specific intelligence. If at all possible, the engineers who conduct the reconnaissance should be trained in this task as well as cross trained with the maneuver unit's scout platoon. The platoon leader must ensure that his designated reconnaissance element links up with the scout platoon.

TYPES

The general types of engineer reconnaissance missions are—

- Technical, which—
  - Is conducted to collect engineer-specific technical data.

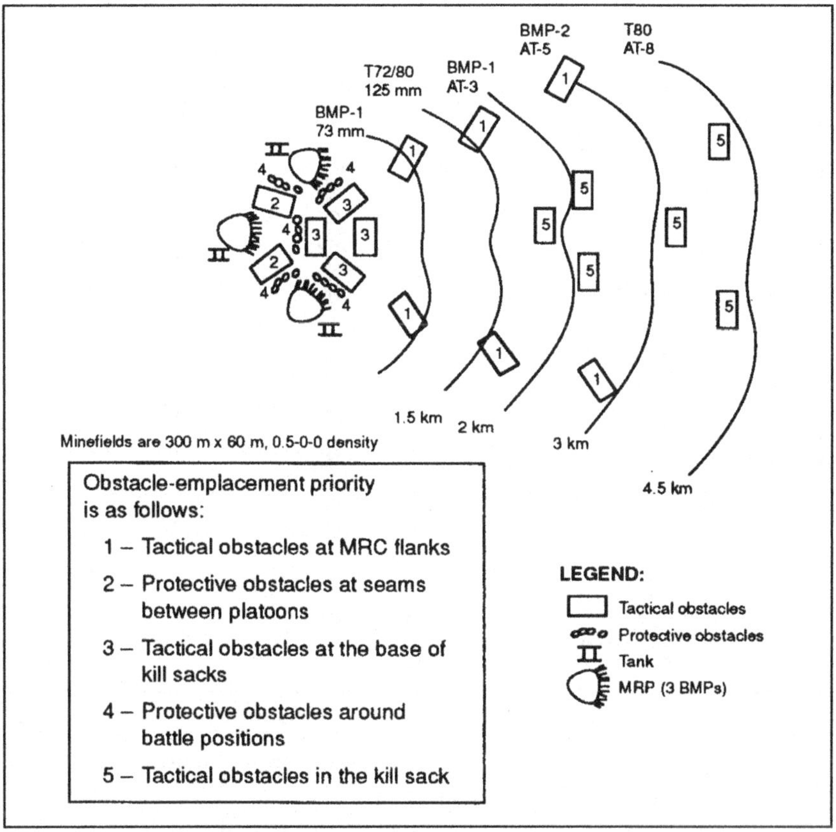

Figure 1-1. Motorized rifle company (MRC) defense template

- Is conducted to perform a specified task from higher HQ in a Level 1 threat environment.
- Involves (almost always) engineers only and is usually conducted behind the forward line of own troops/forward edge of the battle area (FLOT/FEBA).

- Tactical, which—
  - Is conducted to collect engineer-specific information requirements on an enemy point, area target, or attack axis or a route.

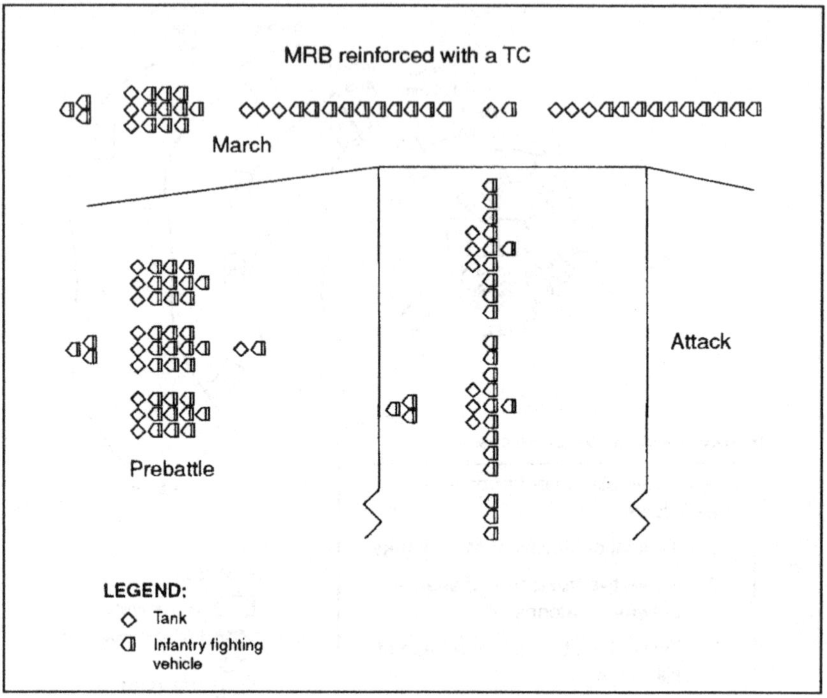

Figure 1-2. Typical threat formations

- Is conducted to support the maneuver plan as an implied task.
- Is conducted in a Level 3 threat environment.
- Requires engineer participation as part of a CA effort.
- Is usually conducted forward of the FLOT/FEBA.

PLANNING

Reconnaissance is a CA effort. The engineers either augment the existing reconnaissance element or conduct their own reconnaissance as part of the consolidated reconnaissance and surveillance (R&S) plan. Engineer reconnaissance must be focused on a specific target, and the reconnaissance element should be given specific intelligence requirements.

In heavy forces, the engineer company commander provides the reconnaissance target and does the majority of the engineer reconnaissance planning. However, in light forces, the engineer platoon leader does the engineer reconnaissance planning.

Platoon reconnaissance planning should include—

- A warning order (WO) based on brigade and TF planning, which includes—
  - Tentative target locations.
  - Type of reconnaissance or target (see *Figure 1-3)*.
  - Specific information, equipment, and coordination requirements.
- The movement of reconnaissance assets to effect the linkup.
- The maximum time for leaders to prepare for the reconnaissance mission.
- The assignment of detailed reconnaissance routes.
- The actions to be taken on contact.

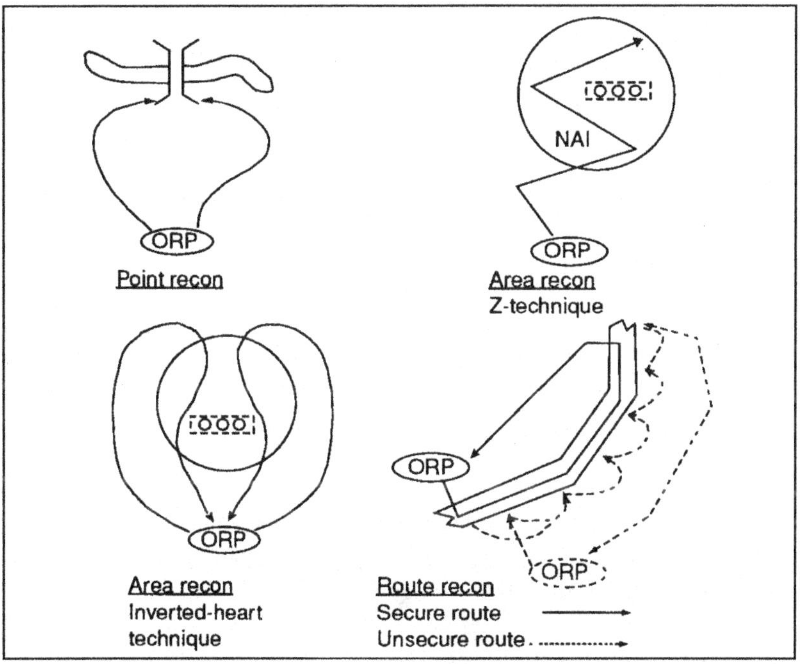

Figure 1-3. Recon techniques

- Consolidation instructions.
- Specific reporting requirements.

MOUNTED AND DISMOUNTED PATROLLING

Patrolling is a CA operation. Engineers are normally part of a larger unit's patrol. When patrolling mounted—

- Position all reconnaissance assets forward.
- Use aggressive and continuous patrols.
- Use area and zone reconnaissance.
- Allow enough time for a thorough reconnaissance.
- Use stealth, not combat power, for reconnaissance.
- Modify formations to overwatch the point element.
- Use terrain driving techniques; use draws and folds in the terrain to mask movement.
- Use night, fog, dust, and storms to conceal movement.
- Reduce speeds to avoid dust signatures.
- Travel at the highest tactical speeds, and place artillery fires on expected enemy positions to avoid long-range fires.
- Keep vehicles dispersed.

When patrolling dismounted—

- Reconnoiter all danger areas, choke points, and critical terrain features.
- Expect contact with enemy dismounted scouts.
- Use darkness, storms, and periods of limited visibility to infiltrate enemy positions.
- Use binoculars or night-vision devices at night with the illumination from ambient light (ensure that laser protection is maintained).
- Use artillery illumination and spotting rounds on reconnaissance targets to assist in land navigation and orientation.

CHAPTER 2
Maneuver

Engineer support to maneuver units requires detailed planning. Engineers may be part of the maneuver unit's formations and, if so, will provide the combat multiplier of engineer effort. *FM 90-13-1* is the primary doctrinal manual for breach planning. *FMs 20-32 and 90-7* are the primary doctrinal manuals for obstacle planning and construction. These references provide the technical information required to plan engineer support for offensive and defensive maneuvers.

The platoon leader should have an understanding of the tactical decision-making process, troop-leading procedures (TLP), and orders preparation. *Figure 2-1, page 2-2,* shows the relationship between TLP, the estimate process, and the intelligence preparation of the battlefield (IPB) process.

In addition, the platoon leader must advise the maneuver commander on the military aspects of the terrain since he is the terrain expert. Normally, the platoon leader only has a map to prepare his terrain analysis. However, he should request terrain-analysis products from his parent engineer company. In both light and armored units, the engineer company has access to terrain-analysis and terrain-visualization products or the ability to develop them. These products greatly aid platoon-level planning.

OFFENSE

The platoon leader is the key engineer expert to his supported unit. He must make several key decisions during the offense. He recommends where critical breach assets are located and which ones are best based on the situation. For example, the supporting engineer platoon leader, in an armored unit, would recommend where and how the M1 tank's plow would be used; or a light platoon leader could recommend to the TF commander a covert breach based on his terrain analysis. In both cases, the engineer platoon leader assists the maneuver commander with his planning. For more information on breaching, see *FM 90-13-1*. The following considerations are specific to the offense:

- Using a highly mobile engineer force, well forward and integrated into maneuver formations, to maintain the momentum of the attack.

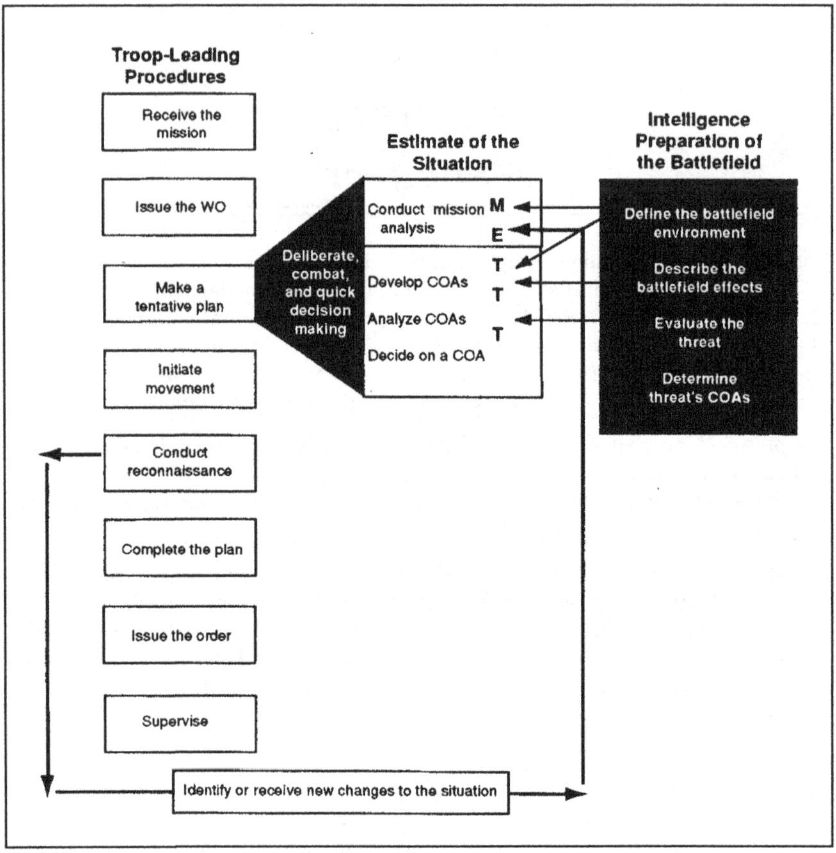

Figure 2-1. Relationship between TLP, the estimate process, and the IPB process

- Arranging for lane turnover between the forward breaching force and follow-on engineers for lane improvement and obstacle clearance.
- Ensuring that marking material and the standing operating procedure (SOP) for lane marking are coordinated between passing units.
- Ensuring that critical engineer equipment, such as armored vehicle-launched bridges (AVLBs), follow-on tactical bridging, lift capability for

line-charge reloading, and lane-marking materials to replenish marking systems, is available.

- Planning situational-obstacle capability for flank security. Scatterable mine (SCATMINE) systems work well in this application.
- Planning tactical-obstacle capability for protection against counterattacks (CATKs). Artillery- and aircraft-delivered family of scatterable mines (FASCAM) are planned for quick emplacement. Ground FASCAM systems are planned for prolonged defenses.
- Ensuring that Class IV/Class V barrier materials are available for a transition to the defense.
- Removing curbs from the AVLB if M1 tanks with rollers are to cross. The roller's dog bone should be raised.

DEFENSE

As with the offense, the platoon leader recommends to the maneuver commander where engineer assets are best used. It is critical that the platoon leader be an integral part of obstacle planning and integration. For more information on obstacle integration, see *FM 90-7*. The platoon leader should also coordinate and develop the obstacle-protection plan. It is critical that the obstacles are protected against enemy breaching attempts, especially covert attempts. The platoon leader should coordinate the inspection of the obstacles that his platoon built before the expected enemy attack. This will ensure that the enemy has not covertly breached and, if so, allows for the obstacles to be repaired before the enemy's main attack. The following considerations are specific to the defense:

- Understanding the commander's intent for obstacles.
- Ensuring that adequate transportation assets are available to haul Class IV/Class V supplies forward (see *Table 2-1, page 2-4*).
- Identifying critical engineer tasks early. Terrain preparation requires time for completion. Engineers must not remain idle while planning is in progress. If possible, mortars, air-defense artillery (ADA), fire-support-team (FIST) vehicles, CSS assets, and the tactical operations center (TOC) should be dug in while direct-fire systems are being positioned. Typically, these systems can be fortified between issuing the WO and completing the direct-fire systems' positioning.
- Practicing operations security (OPSEC) measures to prevent premature disclosure of the defense.

Table 2-1. Class IV/Class V haul capacities

Vehicle	Concertina Wire	M15 AT Mine	M21 AT Mine	M16 AP Mine	M14 AP Mine	MOPMS Mine	Flipper Mine	Volcano Mine	MICLIC Reload
HMMWV, M998 2,500 lb 215 cu ft	2	51	27	55	56	15	11	1	*1
2 1/2 -ton truck 5,000 lb 443 cu ft	4	102	55	111	113	30	23	2	1
5-ton truck 10,000 lb 488 cu ft	7	204	109	222	227	61	46	5	3
5-ton dump truck 10,000 lb ***135/291 cu ft	2/4	112/204	32/70	168/222	71/153	23/51	39/46	3/5	2/3
20-ton dump truck 40,000 lb 754 cu ft	11	628	179	888	443	132	184	20	11
HEMTT truck 20,000 lb 540 cu ft	8	408	128	444	317	94	92	10	7
12-ton S&P 24,000 lb 875 cu ft	13	489	208	533	514	148	110	12	9
40-ton lowboy 80,000 lb 1,760 cu ft	27	1,466	419	1,777	1,035	308	368	43	27
M54812,000 lb 529 cu ft	8	244	125	266	272	74	55	6	4
#Mines / wt/lb — Cube cu ft	40/ 1,180 — 64	1/ 49 — 1.2	4/ 91 — 4.1	4/ 45 — .8	90/ 44 — 1.9	21/ 162 — 5.7	40/ 217 — 3.4	240/ 1,850 — 37.6	** 2,656 — 64.8

for concertina = bundles; 1 bundle = 40 rolls * Overloads vehicle **Line charge + rocket ***Without/with sideboards

- Ensuring that the engineer organization for combat allows for rapid transition to the offense. The reserve must always have a designated force of engineers. Obstacle placement must not interfere with spoiling and CATKs.
- Ensuring that engineer units are not held in reserve but remain committed and work on the commander's priority tasks.

PLATOON LEADER'S PLANNING

The platoon leader can use *Tables 2-2 and 2-3, pages 2-6 and 2-7,* as a guideline when planning and conducting offensive and defensive operations.

REHEARSAL PRINCIPLES

Regardless of the event or task to be rehearsed or the type of rehearsal used, certain principles are universal for conducting effective rehearsals. They are—

- Supporting the scheme of maneuver and the commander's intent.
- Providing clear tasks/conditions/standards (T/C/S).
- Conducting multiechelon CA rehearsals.
- Determining key participants.
- Enforcing standards/training to standard.
- Using the after-action-review (AAR) process to provide feedback to all participants.
- Complementing the preparation phase.
- Instilling confidence in the plan and in the leaders.

During the command estimate process, use the rehearsal principles. These principles produce the—

- Participant levels.
- Rehearsal technique.
- Initiation of precombat checks (PCCs), precombat inspections (PCIs), and mission-specific drills.

Proper application of the rehearsal principles enables the unit to progress through a "crawl/walk/run" process that ends with mission success (see *Figure 2-2, page 2-8*).

Table 2-2. TF-engineer offense briefing

Briefing Topic	Main Points to Brief
Engineer task organization	• How the platoon will be task-organized • Special equipment that will be task-organized to subordinate units • Times and locations for linkup
Enemy situation	• Obstacle template • Enemy's obstacle intent • Enemy's expected level of defensive preparation • Confirmed intelligence on enemy obstacles and positions (include an analysis of this intelligence) • Enemy employment of FASCAM • Impact of enemy obstacles on friendly maneuver
OBSTINTEL collection effort	• Engineer R&S plan • NAIs critical to OBSTINTEL • How the OBSTINTEL effort supports the scheme of maneuver • Scout, engineer, and maneuver OBSTINTEL responsibilities and relationships
Scheme of engineer operations	• How the platoon will support the scheme of maneuver from the objective to the LD, to include consolidation • Engineer contingency plans and pertinent decision points
TF and company/team actions at obstacles	• In-stride breaching, to include— – Areas where the unit can expect to in-stride breach – Special equipment required – Special instructions • Deliberate breaching, to include— – Transitioning from attack formations to breaching formations – How the fundamentals of breaching will be synchronized – Key events and control measures synchronizing the breach – Composition of and instructions for the support, breach, and assault forces – Scheme of fires supporting the breach – Obscuration plan – Coordinating instructions, command and signals for the breach • Assault breaching, to include— – Allocation of engineers – Special coordinating instructions
Countermobility support to offensive maneuver	• Planned locations for situational obstacles • Triggers for situational-obstacle employment
Coordinating instructions	• Lane marking, breach traffic control, order of march through the lane, specific breaching signals, and breach C^2
Service support	• What support the platoon is receiving from the supported company/team and the parent company, if different from the SOP • Medical evacuation, supplies, and vehicle recovery • Current equipment status if critical to the mission
Unresolved issues	• Issues that the plan does not adequately address

Table 2-3. TF-engineer defense briefing

Briefing Topic	Main Points to Brief
Engineer task organization	• How the platoon will be task-organized • Special equipment that will be task-organized to subordinate units • Times and locations for linkup
Enemy	• Enemy maneuver formations through the sector, to include— – Most probable enemy COA – Place and time the enemy will change formation – Enemy breaching assets and capabilities (include high-priority targets)
Countermobility support	• Scheme of obstacles, to include— – Obstacle zones and restrictions – Obstacle belts and groups and their effects – Desired effect of obstacle groups on enemy maneuver – How the scheme of obstacles support the direct- and indirect-fire plans – Triggers and decision points for situational and reserve obstacles – Location of Class IV/ClassV supply points • Obstacle status reporting, to include— – Completion of obstacles – Barrier-material status • Obstacle-protection and turnover plans
Mobility support	• Where the lanes need to be to support MSRs, company/team supply routes, and reconnaissance passages • Responsibility for lane siting and closure • Lane marking and contact points • Obstacle restrictions that support the TF's mobility • Engineer-support plan for mobility
Survivability support	• Protective-obstacle plan, to include— – Material that the unit allocates – Any engineer assistance required – Protective-obstacle reporting responsibilities • Construction plan for fighting positions, to include— – Any engineer assistance required to construct individual positions – Allocation of engineer equipment and time and the priority for constructing vehicle positions – Linkup points and times with the company/team • Equipment control plan to ensure that maneuver companies know— – How many positions will be constructed by type – When, where, and how much equipment they will receive – How long they will retain the equipment
Engineer time line	• Engineer time line, to include— – Estimated completion time – Enemy trigger to stop work – Withdrawal route or passage – Engineer consolidation plan – Class VI/V supply-point recover or destruction
Coordinating instructions	• Obstacle-siting responsibilities • Reporting times • Class IV/Class V supply-point location, security plan, and responsibility

Table 2-3. TF-engineer defense briefing (continued)

Briefing Topic	Main Points to Brief
Service support	• Critical equipment and material shortages affecting defensive preparations • Medical evacuation plan for engineers • Barrier-material configurations to support obstacle-group construction
Unresolved issues	• Issues that the plan does not adequately address

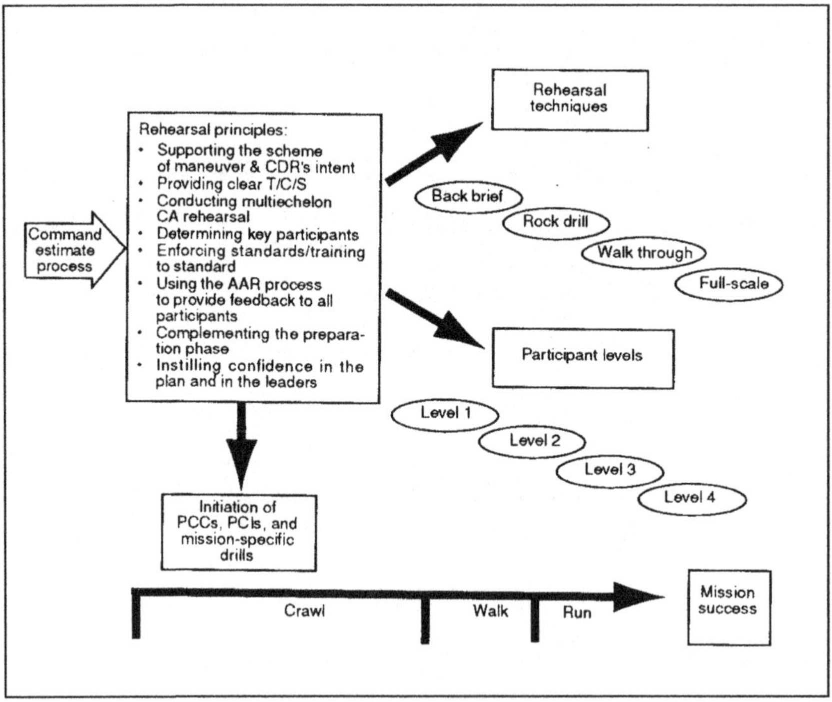

Figure 2-2. Rehearsal planning

PARTICIPANT LEVELS

The participant level details exactly who in the unit is required to attend the rehearsal. A CA-team rehearsal may have four distinct levels. The platoon leader may participate with or without his subordinates in the CA rehearsal. The three levels that apply to a platoon rehearsal follow:

- Level 1 - Squad leaders and platoon HQ.
- Level 2 - Squad leader, team leader, special teams, and vehicle commanders.
- Level 3 - Entire unit.

REHEARSAL TECHNIQUES

Rehearsal techniques fall into four different categories that follow the "crawl/walk/ run" training concept. These categories are—

- Back brief.
- Rock drill.
- Walk through.
- Full-scale.

Each type of rehearsal reflects an increase in mission realism and a corresponding increase in rehearsal payoff. Each technique also increases the realism of the enemy, terrain, subunit actions, and time/distance relationships.

Back Brief

The back-brief rehearsal is an event that occurs after an operation order (OPORD) brief. Subordinate leaders repeat to the commander what he expects them to do and why. They identify all specified and implied tasks, determine their mission-essential tasks, and give their restated mission. During the back brief, they address the—

- Commander's intent.
- Concept of the operation.
- Scheme of maneuver.
- Timing to complete tasks.

During a typical back brief, each leader uses a map or a sand table and explains his mission. The back brief is the quickest of all the rehearsal techniques. It is a leader's tool and is typically the first rehearsal for the unit.

Rock Drill

A rock-drill rehearsal is the acting out of friendly and enemy actions based on the scheme of maneuver and the situation and event templates. Participants rehearse their actions by moving something that represents themselves or their unit, such as rocks or sticks. While acting out the plan, participants talk through their missions, critical tasks, actions, and decisions. Since all participants are simultaneously acting out their part of the scheme of maneuver, leaders can identify problems and disconnects in synchronization more clearly in a rock drill.

Walk Through

A walk-through rehearsal is the acting out of the scheme of maneuver using mounted or dismounted movement based on the assets employed during the attack. Participants should also communicate with the same type of equipment they will use during the action. Participants rehearse by—

- Maneuvering their vehicles (mounted movement) or themselves (dismounted movement).
- Reporting critical actions.
- Making required decisions.

Since the participants are in a more realistic environment, they rehearse the finer aspects of synchronization, C^2, and subunit actions. Aggressive portrayals of enemy actions and reactions are critical in walk-through rehearsals. This threat portrayal is the backbone of increasing the rehearsal realism. This rehearsal technique is more difficult to orchestrate than the back brief or rock drill; however, it is the optimum balance between resource constraints and realism. Successfully mastering a walk-through rehearsal should be the minimum goal for all units.

Full-Scale

During a full-scale rehearsal, participants use real-time mounted and dismounted movement over the actual or similar terrain. Typically, it is conducted at Level 3; however, a mix of participant levels is possible. At least one subunit must participate at Level 3 for a full-scale rehearsal. This type of rehearsal is obviously the most resource intensive, but it provides the most realistic training environment for the unit. It is often used to rehearse the operation plan (OPLAN) or OPORD when time is not an immediate constraint. One must plan extensively to execute full-scale rehearsals correctly and in a manner that does not waste the soldiers' time.

INITIATION OF PCCs, PCIs, AND MISSION-SPECIFIC DRILLS

The subordinate leader conducts PCCs to determine his unit's readiness to execute its assigned tactical mission. These inspections are informal. They are the leader's inspection of mission-critical tasks or areas of special interest to the commander. PCCs should cover—

- Vehicles.
- Weapons systems.

- Soldier's equipment, to include his weapon.
- Mission-critical equipment, such as explosives, detonators, and blasting machines.

The unit commander or leader conducts PCIs to determine the unit's readiness to execute its assigned tactical mission. These inspections may be formal or informal. Formal PCIs are the commander's meticulous and time-consuming inspections of all areas within the unit. They are seldom conducted during combat operations or assembly-area occupation. Informal PCIs are the commander's inspections of particular areas or activities of special interest or concern to him. He may designate the executive officer (XO) or first sergeant (1SG) to perform the PCI. PCIs should cover—

- Vehicles
- Major weapons systems.
- Soldiers.
- Communications equipment.

Mission-specific drills are rehearsals of special combat drills that subordinate units conduct. Special drills may include—

- Breach drills.
- Marking drills.
- Situational obstacles.
- Reserve demolitions.

CHAPTER 3
Mobility and Survivability

Mobility and survivability are the primary tasks of the engineer on the battlefield. The platoon leader requires a detailed understanding of this BOS to succeed. This chapter provides the tactics, techniques, and procedures (TTP) needed for the platoon to conduct mobility, countermobility, and survivability.

MOBILITY

Mobility enables the commander to maneuver tactical units into positions of advantage over the enemy. In the attack, engineers reduce enemy obstacles and fortifications that inhibit maneuver.

Expanded engineer reconnaissance capability is needed to identify routes, existing obstacles, and minefields. Open areas that are level provide good sites for aircraft landing strips; however, in most cases, the soil must be stabilized.

The application of the breaching fundamentals (suppress, obscure, secure, and reduce {SOSR}) and the organization of the force in terms of support, breach, and assault forces are standard. However, open areas may offer greater opportunity to bypass enemy obstacles because of the greater range of mobility afforded by the terrain. Exercise caution when choosing to bypass enemy obstacles, since the bypass may lead the force to the enemy's engagement area (EA). Additionally, expect the enemy to make extensive use of rapid mine-laying techniques that include SCATMINE delivery systems, as well as more conventional methods of mine emplacement.

OBSTACLE BREACHING

The platoon leader must decide where the best breach location is. Never breach where the avenue of approach crosses the obstacle. The enemy is sure to have preplanned artillery there. Pick a location away from obvious breaching sites to reduce the threat of artillery. The platoon leader should place himself where he can see the breach site and effectively command and control his platoon. He must be able to signal his breaching reserve (generally a squad) if he sees the breach failing. Observation of the breach site also reduces reporting requirements on the squad leaders to breach and mark the breach site.

Typically, there is a time lag as the maneuver force suppresses the enemy and obscures the breach site for the breach. The platoon leader should also consider dismounting an element from his lead squad to reconnoiter the best breach location during this tactical pause. This will assist the platoon leader in choosing the best breach site.

Obstacle breaching must be a CA effort. Units are more successful if they conduct—

- Obstacle reconnaissance.
- CA rehearsals.
- Breach planning.

For more information on obstacle breaching, see *FM 90-13-1*.

Breaching Fundamentals

When breaching against a defending enemy, the following fundamentals must be applied to ensure success:

- Suppress. Neutralize or destroy the enemy's weapons with our own fires. (The support force should have a 3:1 advantage over the enemy's weapons overwatching the obstacle.) Consider the effects and characteristics of the enemy's weapons. Ensure the correct focus of direct and indirect fires and the timely shift or lift of fires. Consider what ammunition is required.
- Obscure. Hamper the enemy's observation and target acquisition through the use of smoke (indirect, on board, pyrotechnic), terrain, limited visibility, and stealth.
- Secure. Eliminate enemy interference with obstacle reduction and lane usage. Occupy the obstacle with a friendly force and through the use of effective fires.
- Reduce. Create lanes through, over, or around an obstacle. Ensure that a company/team has at least one lane through an obstacle and that a battalion TF has two lanes. Consider what additional requirements are needed to guide follow-on forces through the obstacle or to mark and reduce the obstacle.

Types of Breaching Operations

When planning a breaching operation, the engineer leader should consider bypassing the obstacle as his first option. However, the enemy's intent may be to force the unit to turn or channelize into a fire sac. This is why a good reconnaissance of a possible bypass is essential. The unit conducting a bypass must have maximum security during the bypass. If bypass operations are not possible, consider the following options:

In-stride Breach. A unit conducts an in-stride breach when it can be successfully executed under the complete C^2 of a task-organized subordinate unit or when the situation is unclear and rapid movement is critical. Subordinate units serve as both the breach and assault forces. The subordinate commanders synchronize the breaching fundamentals. Doctrinally, TFs and above conduct in-stride breaching.

Deliberate Breach. A unit conducts a deliberate breach when the force-allocation ratios for support, breach, and assault forces are beyond the capability of a task-organized subordinate unit. One or more subordinate units are specifically tasked to perform the role of support, breach, and assault forces. Synchronization is critical and the principle of mass drives task organization.

Assault Breach. A unit conducts an assault breach during the penetration and destruction phases of actions on the objective to destroy an enemy that has had time to emplace protective obstacles. For the assault breach, a rule of thumb is one lane per assaulting platoon. Assault lanes are generally only footpaths until the objective is secured. Mounted assault lanes require the same effort as tactical vehicle lanes.

Covert Breach. A unit conducts a covert breach during limited visibility when surprise is essential. The covert breach relies on stealth; quiet, manual lane-reduction techniques; and dismounted maneuver. It can be employed to breach tactical obstacles or protective obstacles (more difficult). It requires the same combat ratios as other breaches for suppression and security, and these points should be considered during the planning process.

Breach Planning

The following are considered when developing a breaching plan (see *Figures 3-1 and 3-2, pages 3-4 and 3-5*):

- Reverse planning begins with actions on the objective.
- Actions on the objective drive the size of the assault force and determine the number and location of lanes to be breached.
- Lane requirements and the type of obstacle drive the allocation of mobility assets to the breach force.
- Ability of the enemy's infantry to interfere with the breach determines whether to secure the breaching site by force or by fire.
- Ability of the enemy to mass fires at the breaching site determines the amount of suppression required and the size of the support force.

The platoon leader must plan for actions after the breach is complete. Generally, the platoon is responsible for beaching, marking, and providing guides for the lane.

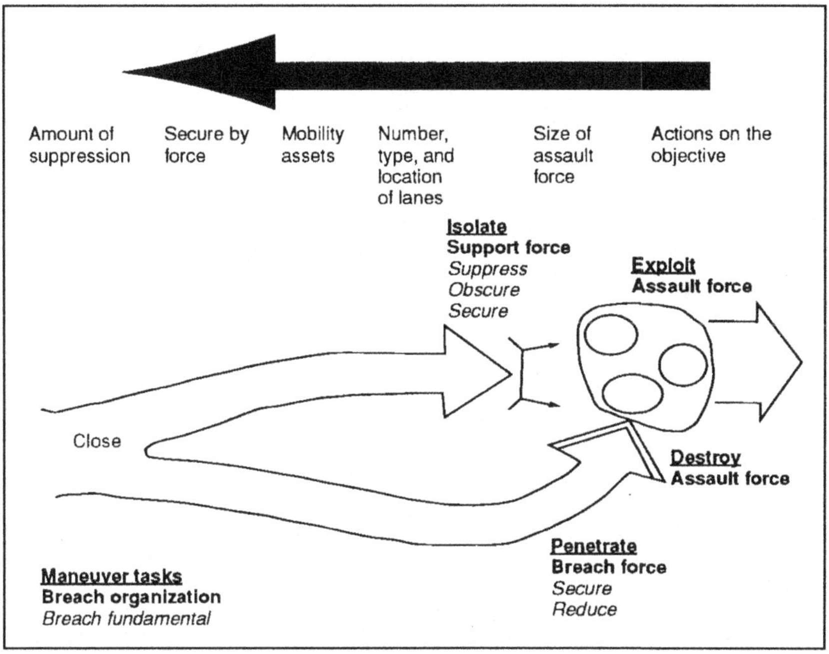

Figure 3-1. Breach planning

Normally, the platoon remains at the breach location and continues to improve the breached lane. The platoon must be prepared to reopen the lane if the enemy closes it with artillery-delivered mines. The platoon leader should also plan for vehicles that are stopped in the lane. The platoon may be forced to create an additional lane around the stopped vehicle or extract it from the breached lane.

Breach TF Organization

The breach TF is organized into three elements:

- Support force—eliminates enemy interference with the breach, isolates the objective area, and destroys enemy weapons that are able to fire on the breach force.
- Breach force—creates lanes in and through the obstacle. While creating the lane through the obstacle, it must also locally secure the breaching site and pass the assault force through the breach.

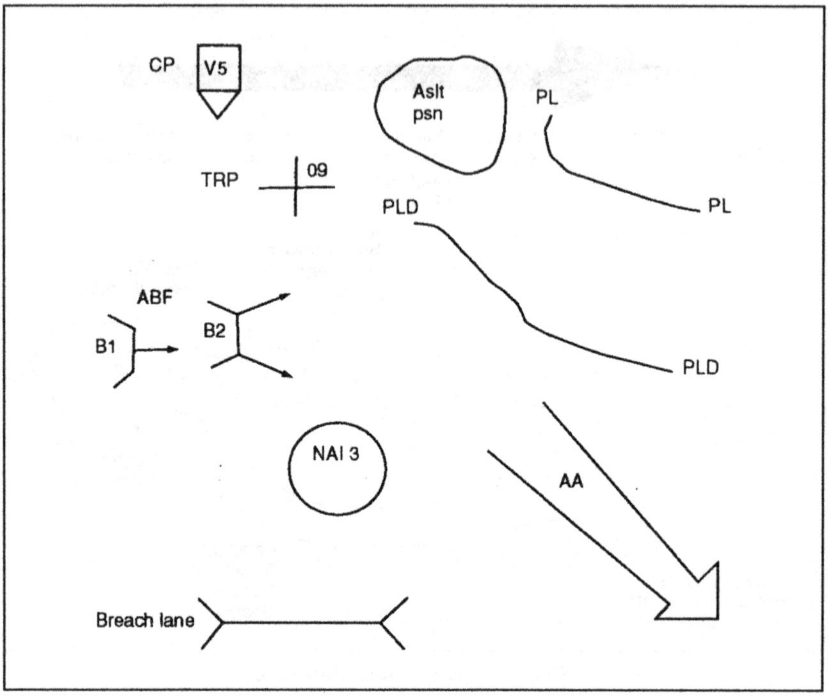

Figure 3-2. Breach control measures

- Assault force—destroys or dislodges the enemy force on the far side of the obstacle and seizes the initial foothold on the objective.

Obscurants

To obscure friendly activities and movement, units can use artillery- or mortar-delivered smoke and/or white phosphorous (WP) rounds, generators, smoke pots and M203 grenades, chemical units, and burning diesel. Obscurants—

- Should be preplanned and coordinated with the company commander and the supported TF.
- Can be used for deception.
- Can be combined with supporting fires.
- Can be used over large areas.

LANE MARKING

The critical components to any lane-marking system are the lane-marking patterns and devices. Lane-marking devices are not standardized Army-wide. *Figure 3-3* shows examples of lane-marking devices. *Table 3-1, page 3-8,* can be used as a guide to choose lane-marking devices. Lane-marking patterns have been standardized Army-wide. Standard breach lane-marking patterns are explained in *FM 90-13-1, Appendix E*. The following are lane markers and their use:

- Entrance markers—indicate the start and the width of a reduced lane. They must be visually different than handrail markers. Entrance markers are placed a minimum of 4.5 meters apart (1 meter for dismounted).
- Handrail markers—define the lane path and indicate the limits of the lane width. Mark, as a minimum, the left handrail.
- Exit markers—indicate the far-side limit of reduced lanes. Exit markers must be visually different from handrail markers but may be the same as entrance markers.
- Entrance-funnel markers—augment the entrance markers. They assist the small-unit commander in guiding the lead unit of his combat column formation.
- Final-approach marker—is a highly visible marker that augments the visual signature of entrance-funnel markers. It provides the assault-force commander with a highly visible reference point toward which to maneuver his formation.
- Far-recognition markers—are highly visible markers located between the final-approach marker and the friendly unit. They are primarily used when passing battalion-sized forces through a lane where distance, visibility, or terrain does not allow the passing force direct observation of the final-approach marker. When possible, far-recognition markers should be different from the final- approach marker.
- Traffic-control post (TCP) or guides—are a two-man team with communications means that assists the commander in controlling the movement of forces. When possible, military police (MP) should man TCPs. However, the breach force should plan to man TCPs until relieved.

The standard levels of lane marking for breach lanes and bypasses are initial, intermediate, and full (see *Figure 3-4, page 3-9*). The following are lane-marking patterns and their use:

- Initial lane-marking pattern—is emplaced by the breach force immediately after the lane is reduced and proofed. It is a signal to the assault force that the lane is ready for traffic to pass through.

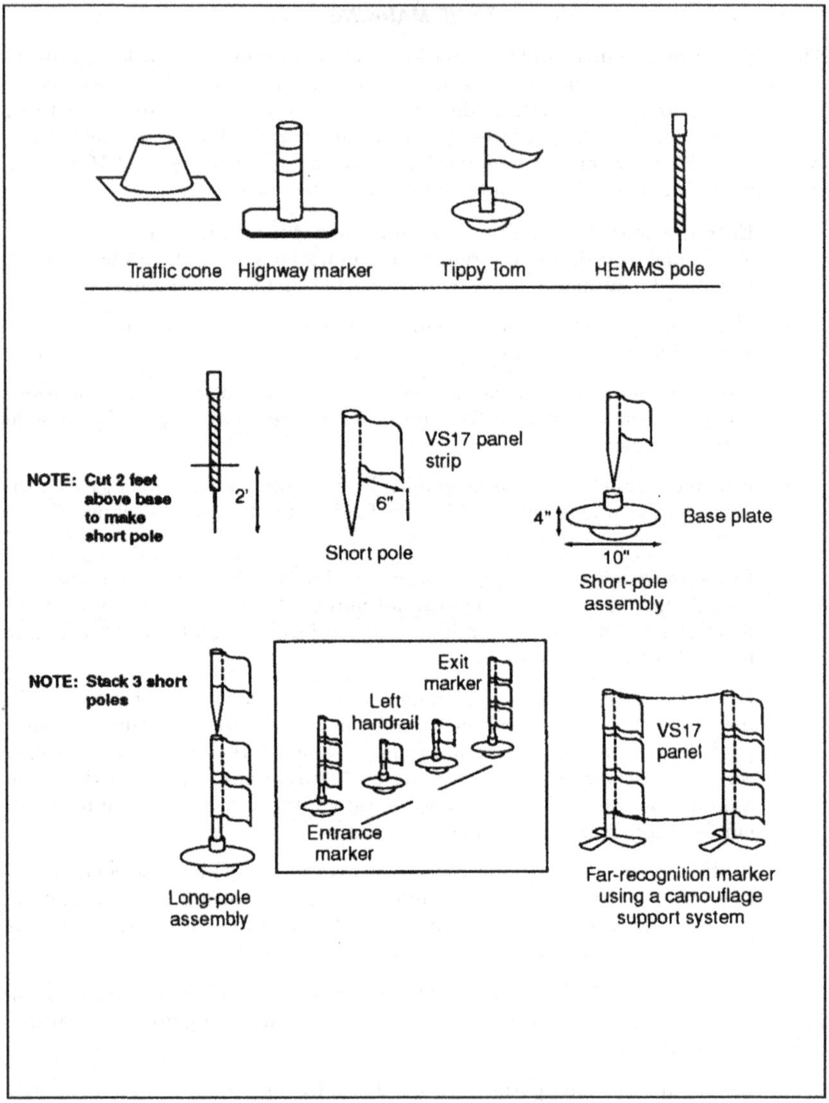

Figure 3-3. Lane-marking devices

Table 3-1. Breach lane marking

Markers	Mounted Forces	Dismounted Forces
Handrail and funnel	They— • Can be seen by the vehicle commander and driver buttoned up from 50 meters. • Are quick to emplace, minimizing soldier exposure.	They— • Are visible to a prone dismounted soldier from 50 meters. • Are lightweight and easy to emplace.
Entrance and exit	They— • Can be seen from 100 meters. • Are visually different from handrail and funnel markers. • Are easy to emplace and are man portable.	They— • Are visible from 50 meters. • Are visually different from handrail and funnel markers. • Are easy to emplace and are man portable.
Final approach and far recognition	They— • Can be seen from 500 meters. • Are visually different from each other. • Can be altered to facilitate traffic control through multiple lanes.	They— • Are visible from 100 meters. • Are visually different from each other. • Can be altered to facilitate traffic control through multiple lanes.

- Intermediate lane-marking pattern—is used for the commitment of larger combat forces who are unable to directly observe the breach or the rearward passage of sustainment traffic. It builds on the initial lane-marking pattern by adding right handrail markers, exit-funnel markers, far-recognition markers, and a far-side final-approach marker.
- Full lane-marking pattern—is usually not part of an initial breaching operation. It is a lane that will support uninterrupted two-way traffic.

Table 3-2, page 3-10, provides information as to who marks the various types of lanes, when they are to mark the lanes, and what markers they are to use. Lane marking is normally synchronized during the CA rehearsal to ensure that all key leaders understand the marking procedure and standard being employed.

COUNTERMOBILITY

Countermobility operations are conducted to augment natural terrain with obstacle systems according to the commander's concept. This adds depth to the battle in

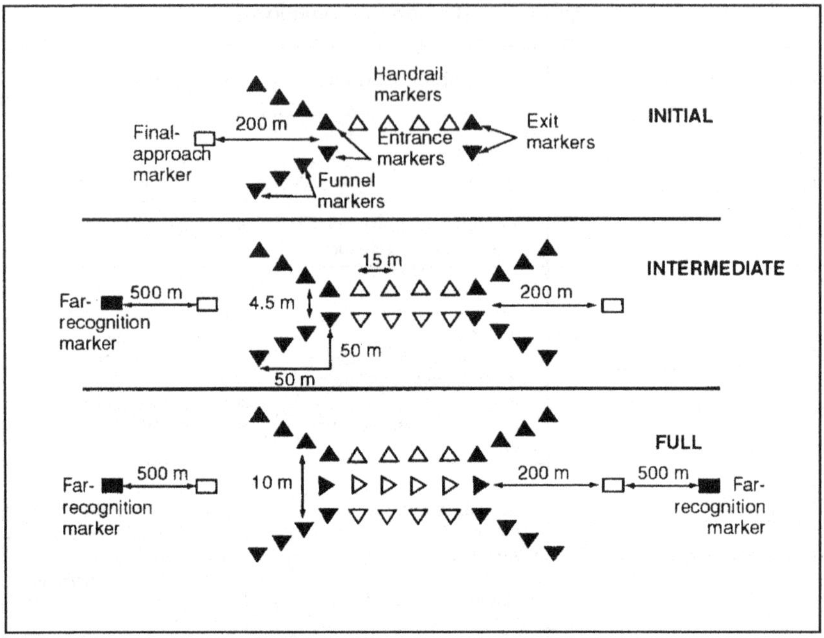

Figure 3-4. Lane-marking patterns

space and time by attacking the enemy's ability to maneuver its forces. With the enemy's movement impeded (disrupted, turned, fixed, or blocked), it is vulnerable to our forces. Engineers advise the commander on the best means to reinforce natural obstacles (terrain) and to emplace tactical obstacles that support his plan. Minefields, wire, antitank ditches (ATDs) or berms, road craters, bridge demolition, and all other traditional obstacles may be effective. For expedient obstacles, units should consider employing local materials.

ATD or berms require extensive preparation but are very effective when properly prepared. Many areas have irrigation ditches that can be used tactically. However, in sandy areas, ditches can easily be filled in. Obstacles should be planned to protect the unit's flanks during offensive operations.

All obstacles can be situational obstacles. Situational obstacles are obstacle resources held in reserve. They have a "be-prepared" mission but not an "on-order mission". Situational obstacles should be considered during planning to give the maneuver commander a flexible response to enemy or friendly maneuver.

Table 3-2. Lane-marking level—unit responsible, trigger events, and lane markers

Breach Type	Initial	Intermediate	Full
	Unit Responsible		
Deliberate	TF breach force	TF breach force	Brigade
Covert	TF breach force	TF breach force	Brigade
In-stride	Breach company/ team	TF mobility reserve	Brigade
Assault	Assault platoon	TF assault force	NA
	Trigger Events		
	When— • Lanes are reduced • Passing platoon- or company-sized forces	When passing— • Battalion- or company-sized forces • Forces that can-not see the lane • TF combat trains	When— • Passing brigade- or battalion-sized forces • Situation requires uninterrupted sus-tainment traffic
	Lane Markers		
	Entrance/exit Left handrail Entrance funnel Final approach	The following mark-ers are added: • Right handrail. • Exit funnel. • Far-side final approach. • Far recognition. • Guides/TCPs.	Lane width is expanded to 10 meters. Existing markers are adjusted. Far-side-recognition markers and guides/TCPs are added.

OBSTACLE PLANNING

To assist the TF commander in the defense, the TF engineer must integrate obstacle planning throughout the TF sector and coordinate with adjacent TFs according to the overall engineer plan.

Each echelon has its own obstacle-control measures for directing the location of obstacles on the ground. The division commander and his staff use obstacle zones to control obstacles to support future operations. Within these zones, the brigades plan obstacle belts, and the battalion TF plans obstacle groups. The engineer will then emplace each obstacle by doctrinal obstacle norms to support the commander.

Protective Obstacles

They are designed to repel the enemy's assault by emplacing them around a unit's defensive position (a survivability task).

Tactical Obstacles

They are designed and emplaced to (see *Figure 3-5*)—

- Disrupt, which—
  - Breaks up C^2.
  - Frustrates the enemy's timing.
  - Interrupts formations or attacks the enemy when they are massing into formations.
  - Coerces the enemy into our lines of fire.

Obstacle-Effect Graphic	Application	Examples Conveying Intent	Resource Factor[1]
Disrupt 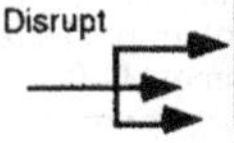	Short arrow indicates where enemy is attacked by obstacles. Long arrows indicate where bypass is allowed and attacked by fires.	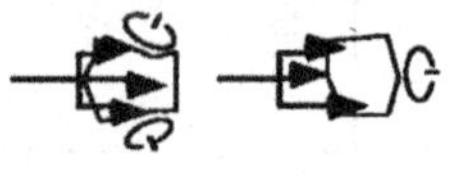	0.5
Turn 	Heel of arrow is anchor point. Direction of arrow indicates desired direction of turn.		1.2
Fix 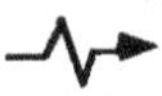	Irregular part of arrow indicates where enemy advance is slowed by obstacles.	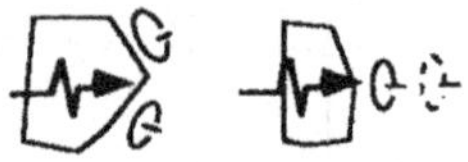	1.0
Block 	The ends of the vertical line indicate the limit of enemy advance. The ends of the vertical line also indicate where obstacles tie in to severely restricted terrain.	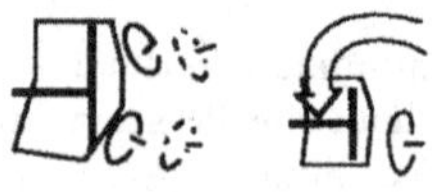	2.4

[1] See *FM 90-7, Appendix C*, for an explanation of how to use the resource factor.

Direction of Enemy Attack

Figure 3-5. Obstacle-effect graphics

- Fix, which—
  - Slows the enemy's advance within a specified area.
  - Causes the enemy to breach repeatedly.
- Turn, which manipulates enemy maneuver in a desired direction.
- Block, which—
  - Stops the enemy along a particular avenue of approach.
  - Defeats the enemy's breaching effort.

Figure 3-6 shows how to employ conventional row minefields to obtain the desired obstacle effects and the planning figures for standard minefields by obstacle effect.

Situational Obstacles

Situational obstacles are "triggered" by an event or situation. Once the general location of the obstacle has been determined, it must be refined to better enhance its effect on the enemy. When planning, preparing, and executing situational obstacles (see *FM 90-7*)—

- Identify the need.
- Plan for appropriate resources.
- Plan the obstacle.
- Integrate the obstacle with friendly fires.
- Identify obstacle execution triggers.
- Withhold execution of the obstacle until it is needed.

Situational obstacles can be used to—

- Attack an enemy's vulnerability.
- Exploit success.
- Separate follow-on enemy forces.
- Provide flank protection.

Tables 3-3 and 3-4, pages 3-14 and 3-15, provide some characteristics of conventional and scatterable mines and are useful in planning minefields. When emplacing standard disrupt, fix, turn, and block minefields, see *Figures 3-7 through 3-12, pages 3-16 through 3-21, and Tables 3-5 through 3-8, pages 3-22 and 3-23,* for information on the employment of SCATMINE systems. FASCAM systems include the air/ground Volcano, Modular Pack Mine System (MOPMS), Flipper, area denial artillery munition (ADAM), and remote antiarmor mine (RAAM).

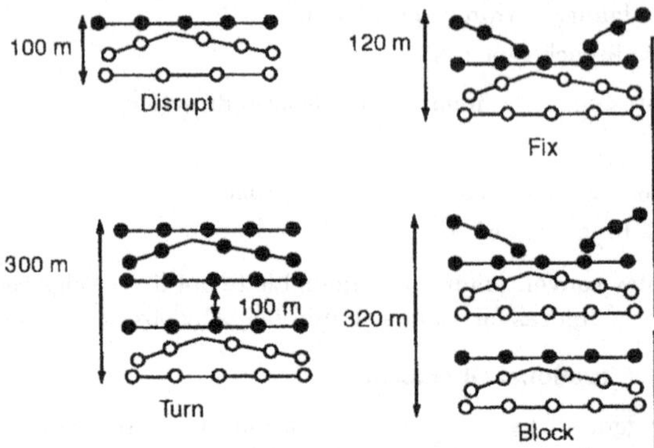

Mines spaced 6 m apart
Rows spaced 50 m apart
Minefield frontage
Disrupt: 250 m
Turn: 500 m
Fix: 250 m
Block: 500 m
o Track-width fuzed mine
● Full-width fuzed mine

AP mine employment

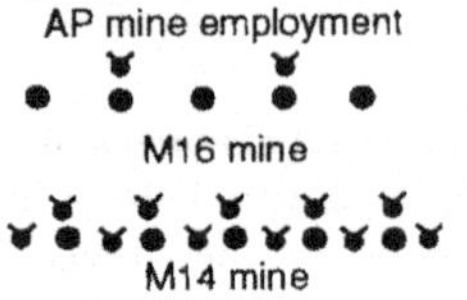

AP mines are—

- Used in block minefields only.
- Used in any two rows of full-width AT mines.
- Patterned based on AP mine type.

	Disrupt	Turn	Fix	Block
Frontage (m)	250	500	250	500
Depth (m)	100	300	120	320
AT full (# rows)	1	4	1	4
AT track (# rows)	2	2	2	2
AP (# rows)	0	0	0	2
IOE (yes or no)	No	No	Yes	Yes
AHD (yes or no)	Yes	No	No	Yes
Plt hrs reqd	1.5	3.5	1.5	5.0
Track-width AT mines	84	168	84	168
Full-width AT mines	42	336	63	378
AP mines M16	0	0	0	84
AP mines M14	0	0	0	500
Density AT	0.5	1.0	0.6	1.1
Density AP M16/M14	0	0	0	0.17/1.0

Figure 3-6. Minefield norms

Table 3-3. Conventional mines

Mine	Arming Time	Fuzing	Warhead	AHD	Mine Weight	Mines per 5-Ton Dump	Remarks
M14	NA	Pressure	Blast	No	3.3 oz	6,480	
M16A1	NA	Pressure Trip wire	Bounding fragment	Yes	8 lb	672	
M18A1	NA		Directional fragment	No	3.5 lb	1,782	
M86 (PDM)	50 sec		Bounding fragment	100%	1 lb	—	Self-destruct after 4 hours
M15	NA	Pressure	Blast	Yes	30 lb	90	
M15 w/ M624	NA	Tilt rod	Blast	Yes	30 lb	90	
M21	NA	Tilt rod	Shape charge	Yes	17 lb	192	

OBSTACLE SITING

When determining where to emplace obstacles, the platoon leader should consider the following:

- Supporting the maneuver commander's intent.
- Supporting the CATK plan.
- Supporting the defensive plan.
- Coordinating obstacle locations with the maneuver element, the fire-support officer (FSO), the Operations and Training Officer (US Army) (S3), and the weapons-systems commander.
- Prioritizing the engineer effort.
- Planning the maneuver unit's responsibility.
- Using observation and fields of fire, cover and concealment, obstacles, key terrain, and avenues of approach (OCOKA).

Company/Team Commanders' Coordination

The platoon leader must ensure that key obstacles are covered by indirect fire. These obstacles become priority targets for the artillery. When coordinating with company/team commanders, the platoon leader should consider the following:

- Determining the obstacle intent.
- Determining the company's/team's tactical purpose.

Table 3-4. Scatterable mines

Mine	Delivery System	Arming Time	Fuzing	Warhead	AHD	Self-Destruct Time	Mine Weight	Mines per 5-Ton Dump
M73	155-mm arty (RAAM)	45 sec 2 min	Magnetic	M-S plate	20%	48 hr	3.8 lb	NA
M70	155-mm arty (RAAM)	45 sec 2 min	Magnetic	M-S plate	20%	4 hr	3.8 lb	NA
M75	GEMSS Flipper	45 min	Magnetic	M-S plate	20%	5 days 15 days	3.8 lb	1,600
BLU 91/B	USAF (Gator)	2 min	Magnetic	M-S plate	No	4 hr 48 hr 15 days	3.8 lb	NA
M76	MOPMS	2 min	Magnetic	M-S plate	No	4 hr (recycle 3 times)	3.8 lb	30 modules (510 mines)
Volcano	Ground/air	2 min	Magnetic	M-S plate	No	4 hr 48 hr	3.8 lb	160 canisters (800 mine)
M72	155-mm arty (ADAM)	45 sec 2 min	Trip wire	Bounding fragment	20%	15 days 48 hr	1.2 lb	36 per M692 projectile
M67	155-mm arty (ADAM)	45 sec 2 min	Trip wire	Bounding fragment	20%	4 hr	1.2 lb	38 per M731 projectile
M74	GEMSS Flipper	45 min	Trip wire	Blast fragment	20%	5 days 15 days	3.2 lb	1,600
BLU 92/B	USAF (Gator)	2 min	Trip wire	Blast fragment	100%	4 hr 48 hr 15 days	3.2 lb	NA
M77	MOPMS	2 min	Trip wire	Blast fragment	100%	4 hr (recycle up to 15 days)	3.2 lb	30 modules (120 mines)
Volcano	Ground/air	2 min	Trip wire	Blast fragment	100%	4 hr 48 hr 15 days	3.2 lb	160 canister (160 mines)

- Ensuring that the obstacles the maneuver company/team commanders emplace on the battlefield serve as a combat multiplier and enhance the direct-fire plan.
- Providing forward security during obstacle emplacement.
- Conducting obstacle turnover with the maneuver company/team and ensuring that indirect-fire support is provided after the obstacle is finished.
- Rehearsing lane-closure plans.

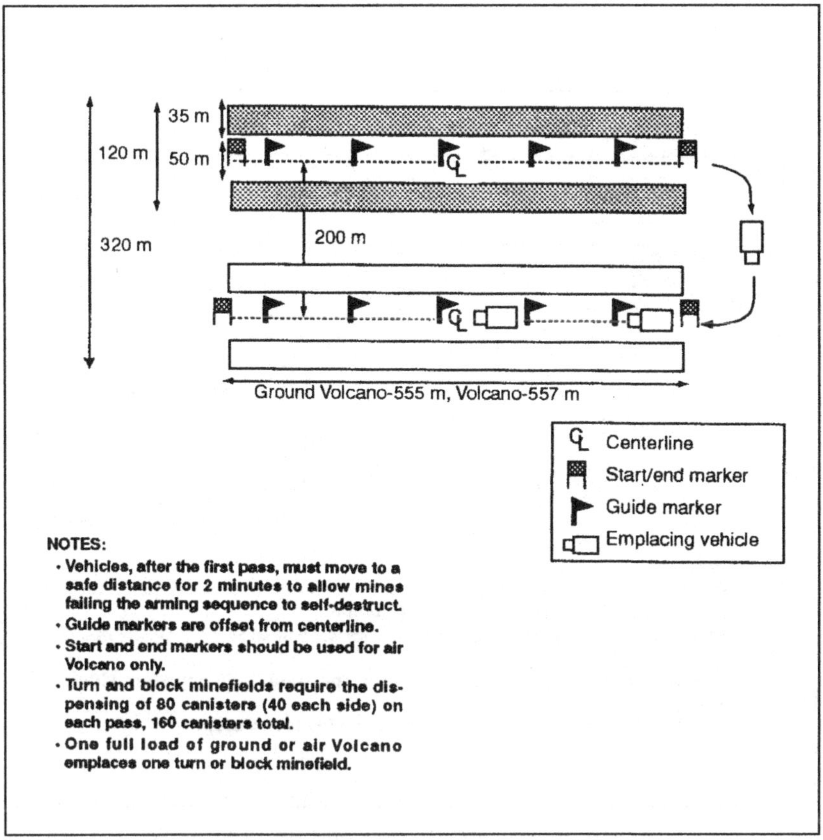

Figure 3-7. Volcano turn and block minefields

Direct-Fire Integration

When integrating obstacles with direct fire, the platoon leader should consider the following:

- Coordinating with the maneuver element overwatching the obstacle.
- Sighting obstacles with weapons.
- Coordinating with the maneuver element for—
  - Forward security.

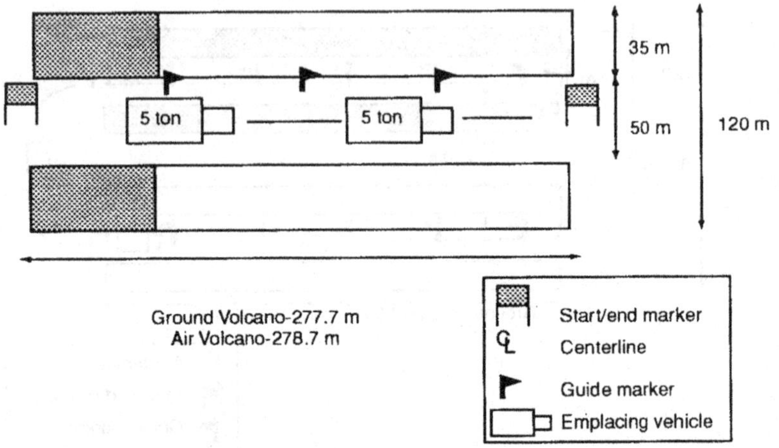

Notes:

- **Guide markers are offset to the left of the centerline to guide the driver and allow the vehicle to remain on the centerline.**
- **Start and end markers should be used for air Volcanos only. Guide markers are not required.**
- **Disrupt and fix minefields require the dispensing of 40 canisters (20 each side).**
- **One full load of air or ground Volcano provides four disrupt or fix minefields.**

Figure 3-8. Volcano disrupt and fix minefields

- Logistics support.
- Obstacle-turnover requirements.

Indirect-Fire Integration

The platoon leader must ensure that key obstacles are covered by indirect fire. These obstacles become priority targets for the artillery. The platoon leader coordinates indirect fire with the company FIST and company/team commander in heavy units. Indirect fire must be coordinated with the FSO and the TF commander if units are light or if the platoon leader is the TF engineer. For more information on

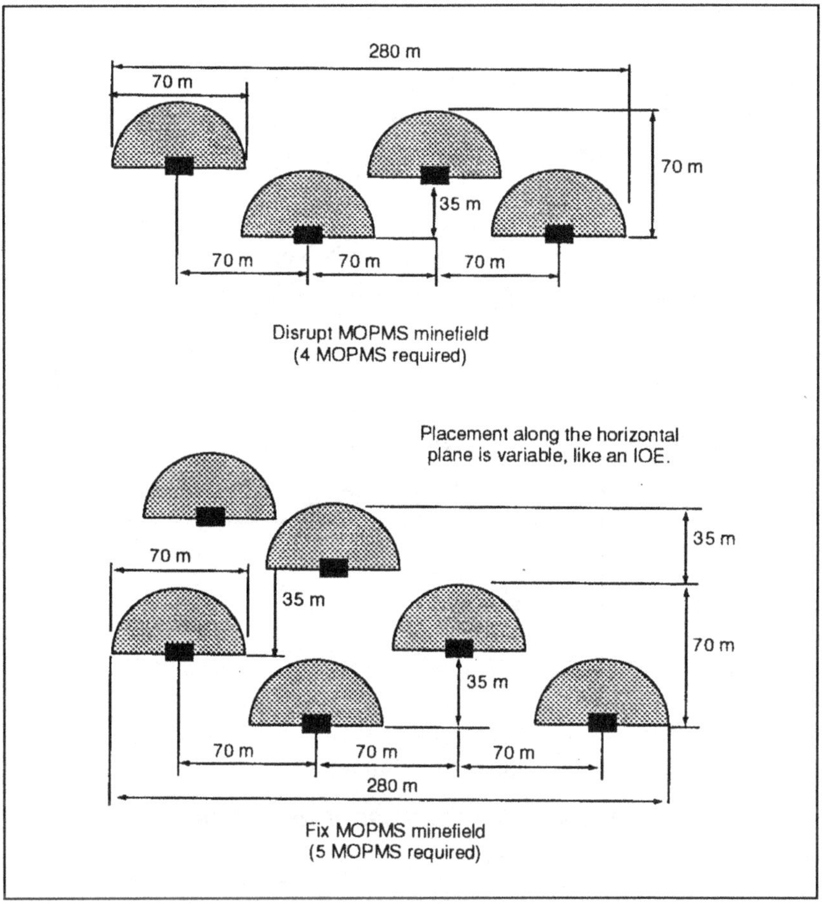

Figure 3-9. MOPMS disrupt and fix minefields

integrating obstacles with indirect fire, see *FM 90-7*. When integrating obstacles with indirect fire, the platoon leader should consider the following:

- Ensuring that there are target reference points (TRPs) on every obstacle.
- Preplanning ADAM and RAAM fires.
- Calling for fire nets (engineers).

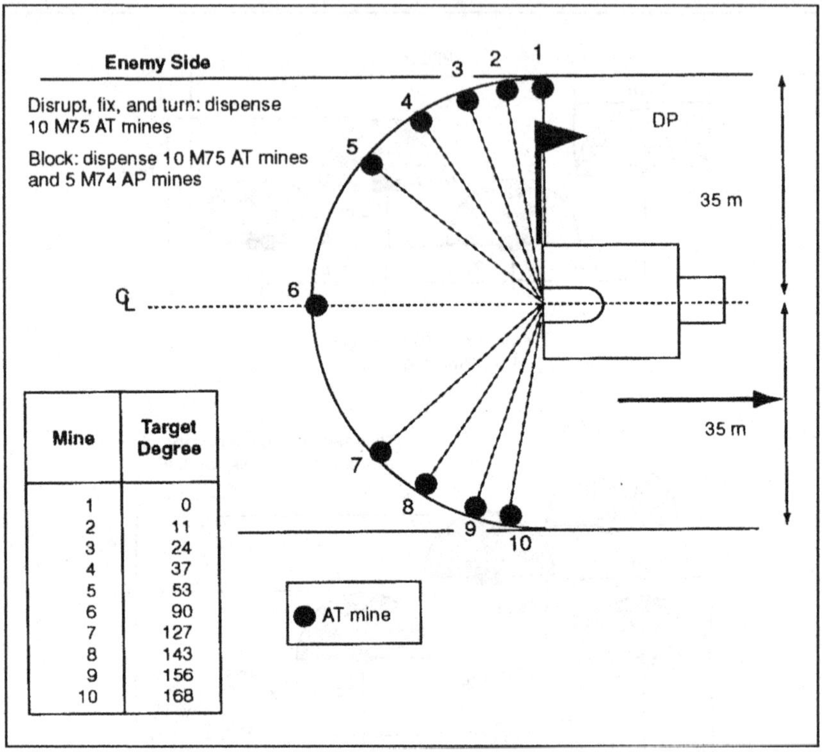

Mine	Target Degree
1	0
2	11
3	24
4	37
5	53
6	90
7	127
8	143
9	156
10	168

Figure 3-10. Flipper TTP

OBSTACLE CAMOUFLAGE (IF REQUIRED)

When camouflaging obstacles, the following should be considered:

- Ensuring that the obstacles emplaced support the intent (sometimes you want the enemy to see the obstacles).
- Using reverse slope siting.
- Burying mines.
- Using night and covert emplacement.
- Minimizing dust signatures from the excavation.

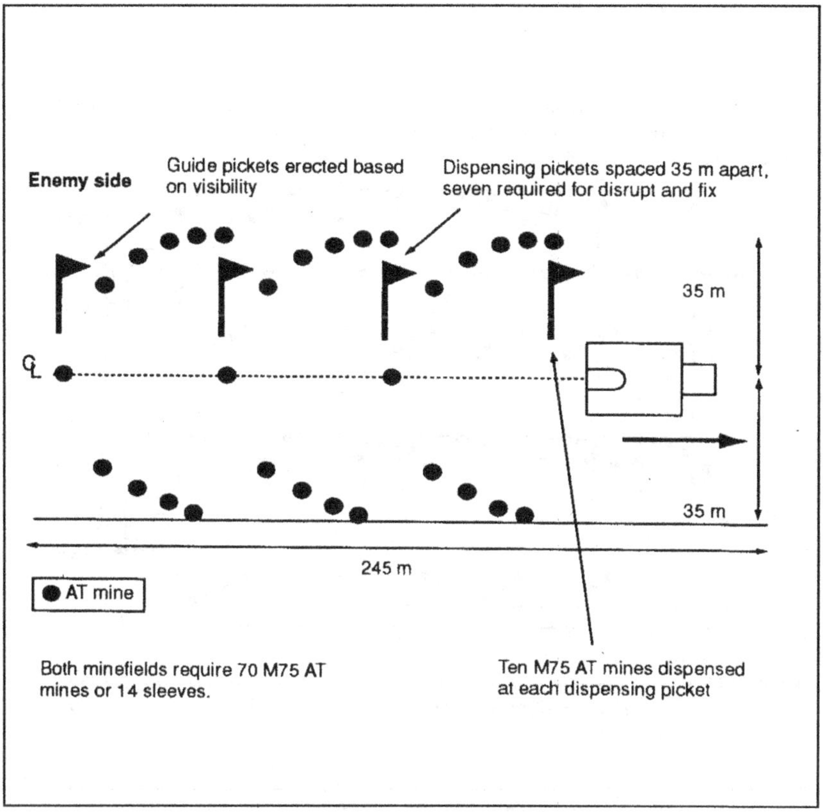

Figure 3-11. Flipper fix and disrupt minefields

OBSTACLE PROTECTION

The following should be considered for the protection of obstacles:

- Denying the enemy covert breaching opportunity.
- Denying the enemy reconnaissance
- Using observation posts (OPs) and counterreconnaissance.
- Displacing forces during low visibility to cover obstacles.

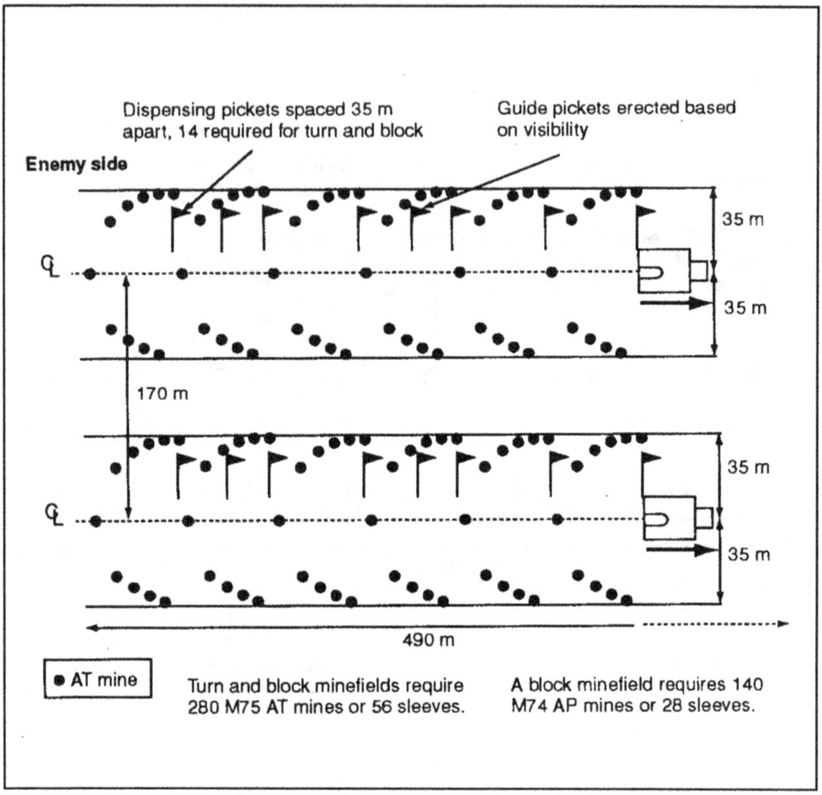

Figure 3-12. Flipper turn and block minefields

- Identifying and destroying the enemy's obstacle breaching equipment and C^2 vehicles.
- Covering the obstacle with direct and indirect fires.
- Patrolling obstacles at night.

SURVIVABILITY

Preparation of fortifications may be difficult. Fortifications in sandy soils often require revetments. It may be impossible to dig in rocky areas, arctic regions, or locations with a high or perched water table. To counter these problems, you may

Table 3-5. Flipper planning factors

Type of Minefield	Depth (Meters)	Front (Meters)	Number of Strips	DPs per Strip	M75 Mines per DP	M74 Mines per DP	Total M75 Mines	Total M74 Mines
Disrupt	70	245	1	7	10	0	70	0
Fix	70	245	1	7	10	0	70	0
Turn	240	490	2	14	10	0	280	0
Block	240	490	2	14	10	5	280	140

Table 3-6. Volcano planning factors

Type of Minefield	Depth (Meters)	Front Ground/Air (Meters)	Number of Strips	Canisters per Strip	Total Canisters	Minefields per Load
Disrupt	120	277/278	1	40 (20 each side)	40	4
Fix	120	277/278	1	40 (20 each side)	40	4
Turn	320	555/557	2	80 (40 each side)	160	1
Block	320	555/557	2	80 (40 each side)	160	1

Table 3-7. RAAM/ADAM planning factors

Type of Minefield	RAAM		ADAM		Width (Meters)	Depth (Meters)
	Area	Linear	Area	Linear		
Disrupt	0.001	0.2	0.0005	0.1	200	200
Turn	0.002	0.9	0.001	0.4	400	400
Fix	0.002	0.4	0.0005	0.1	200	200
Block	0.004	1.6	0.002	0.8	400	400

Table 3-8. Mine dump planning factors

Number of Personnel	Quantity of Mines
2-man team (2 minutes per mine)	25 mines per hour
Squad (8 soldiers)	100 mines per hour
Platoon	300 mines per hour; 3,000 mines per day
Company	6,000 mines per day
NOTE: For planning, soldiers work 10 hours per day.	

have to build up emplacements with existing materials or depressions. *FMs 5-34 and 5-103* provide the standards for dismounted and vehicular field fortifications.

Camouflage is very effective when properly employed. However, you must carefully select patterns and techniques to apply to the local environment. All equipment should have camouflage nets. Other survivability techniques include dispersion and frequent moves.

Direct-fire weapons are more effective from dug-in positions. Logistics areas (brigade support area/division support area (BSA/DSA)) require survivability support. These sites are large, relatively static, and difficult to camouflage. As a result, they are vulnerable to enemy interdiction. Emphasis should be placed on ammunition and fuel dumps.

Digging assets, such as bulldozers, small emplacement excavators (SEEs), or armored combat earthmovers, M9 (ACEs), should be tasked to provide survivability support to these sites. Particular emphasis should be placed on hardening ammunition and fuel-storage locations.

With long-range observation and fields of fire, defensive positions are very vulnerable to offensive fire. This fact, coupled with a lack of natural obstacles, may lead the commander to concentrate the bulk of his engineer effort on survivability positions. Survivability positions enhance the ability of all direct-fire elements to survive indirect-fire preparations and the direct fire from attacking enemy units.

Care should be taken when digging foxholes and tank hide positions in sandy or noncohesive soils since the walls have a tendency to cave in.

PLANNING

Survivability positions are key to surviving any battle on today's modern battlefield. This is especially true where there is a lack of vegetation and natural hiding positions. The TF engineer must know the following information to start planning:

- Number of positions: primary, alternate, and supplementary.
- Type of position: hull-down, turret-down, or hasty.
- Weapons and vehicle priorities.
- Battle-position and EA priorities.
- Unit priorities.

Once the TF engineer determines the number of positions required, then the number of positions that can actually be put in is determined. The engineer unit's capability is based on the following factors:

- Ten-hour days for soldiers.
- Fifteen-hour days for equipment.
- Time available (days).
- Number of blade teams (BTs) available. ACEs, dozers, combat engineer vehicles (CEVs), and bucket loaders can be used to make blade teams. A blade team must include at least one dozer or ACE.
- Work rates.

For work rates, use the planning factor developed for your area of operation. If work rates are unknown, use the following as a guide:

- One hull-down position (HDP) takes 1.5 BT hours.
- One turret-down position (TDP) takes 3.5 BT hours.
- Fifty meters of ATD takes 1 BT hour.

A variety of variables may increase or decrease the time required to dig a vehicle fighting position or ATD. These variables include, but are not limited to, local soils and terrain, operator experience, additional maintenance down time, and the tactical situation. The difference of blade widths may also affect work rates. The M9 ACE has a shorter blade width than the D7 dozer and requires a minimum of two passes to cut the same width.

Survivability assets should be massed on the commander's number one priority. This focuses the survivability effort and allows the TF engineer to monitor construction progress. It also reduces the C^2 burden on the engineer platoon and facilitates rapid transition to changing priorities.

The following equations can be used to determine the number of HDPs and TDPs and the length of ATDs that can be emplaced:

HDP = (# BTs) x (1 HDP/work rate) x (# days) x (15 hrs/day)

TDP = (# BTs) x (1 TDP/work rate) x (# days) x (15 hrs/day)

Meters of ATD = (# BTs) x (ATD m/work rate) x (# days) x (15 hrs/day)

POSITIONS

Coordination between the engineer/equipment operator and the maneuver element is critical to ensure the proper placement of positions. Without effective coordination, much valuable time is wasted redigging positions. The maneuver element, preferably the gunner or tank commander (TC), must tell the engineer/equipment operator the following:

- Type of position (deliberate or hasty).
- Gun target line.
- Sector of fire.
- TRP.

In heavy units, initial vehicle-fighting-position construction starts with the M9 ACE. The ACE has different digging capabilities than the dozer. In many cases, divisional ACEs start with hasty positions, which are then upgraded to improved positions as time allows. Also, the ACEs may start work before corps-level dozers arrive to augment the construction effort. In this situation, the ACEs would dig the hasty positions and the dozers would improve them. The platoon must plan for this augmentation to efficiently integrate corps assets into the construction effort. In heavy units, the company commander must integrate corps assets; however, in light units, the platoon leader must plan for and integrate corps assets.

Figure 3-13, page 3-26, shows the different vehicle fighting positions. Some terrain areas may not accommodate a hull or turret defilade. The use of a modified hasty vehicle-position design or simple marking techniques incorporated with terrain mottling may be required.

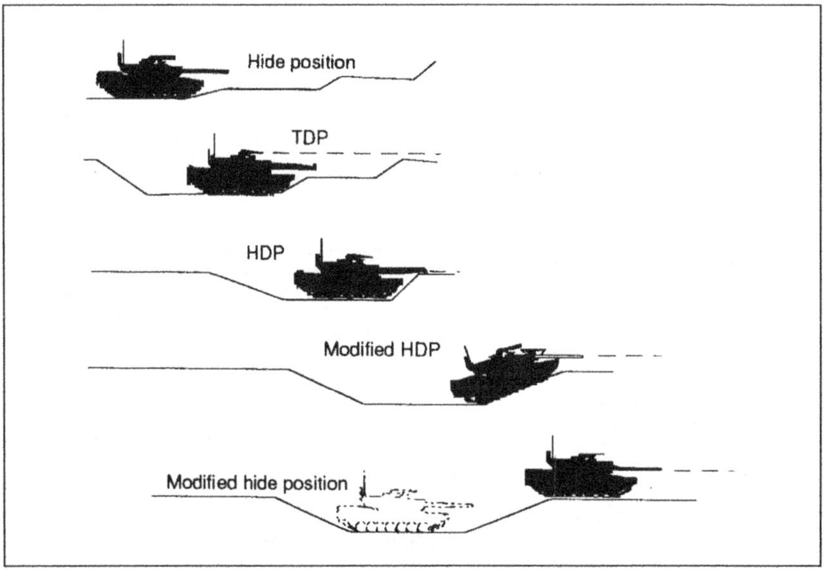

Figure 3-13. Vehicle survivability positions

Each soldier, not just engineers, must also emplace individual and crew fighting positions. Engineers have some limited assets, such as the SEE, that can assist the maneuver commander in accomplishing this mission. Fighting positions are—

- Located to effectively engage the enemy and mutually support each other.
- Designed to allow weapons to be properly fired.
- Constructed to provide adequate cover (see *Table 3-9*).
- Concealed from the enemy.

NUCLEAR, BIOLOGICAL, CHEMICAL (NBC)

Chemical and biological weapons are many countries' preferred weapons of mass destruction. Many of our potential adversaries have an NBC capability. *Figure 3-14 and Tables 3-10 and 3-11, page 3-28,* detail the NBC threat and protective procedures.

Table 3-9. Survivability

Camouflage	Defensive Positions
Soldiers should consider the following when camouflaging: • Breaking up the shape of the object. • Shading the object with natural shadows. • Concealing the shine of the object. • Disrupting the silhouette of the object. • Using folds, gullies, and wadies. • Using reverse-slope positions. • Scaring two to three positions for each vehicle or position. • Using irregular-shaped positions • Building positions a minimum of twice the size of the concealed object. • Exercising noise and light discipline. • Using camouflage for deception.	Soldiers should know and consider the following facts when constructing defensive positions: • Field expedient construction materials, such as corrugated metal, coverts, scrap metal, 40-gallon oil drums, sandbags, and sand-filled ammo boxes, can be used. • Small-arms fire penetrates loose sand 39 inches and sandbag parapets 24 inches. • A minimum of 18 to 24 inches of overhead cover is required to protect against indirect fire. • Revetments are required in sandy soils. • Design details are covered in *FMs 5-34 and 5-103.*

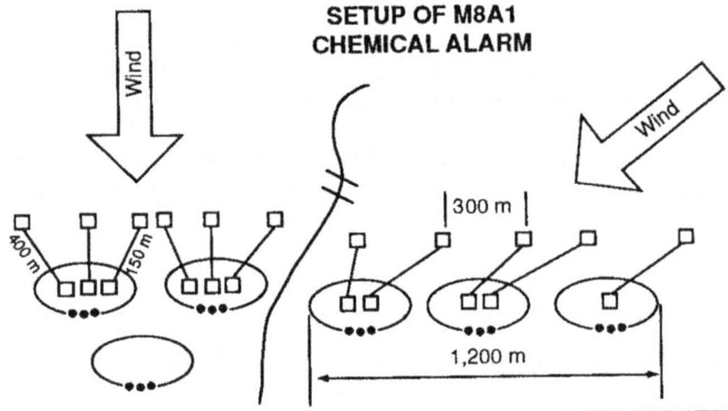

CAM put in operation

1. Remove nozzle's protective cap and place cap onto environmental cap.
2. Place filtered nozzle standoff onto nozzle (do not touch nozzle).
3. Insert nozzle in "H" end of confidence sample for 1 second.
   - Verify that at least 3 bars are displayed.
   - Ensure that the display has no more that 1 bar within 2 minutes.
4. Check in "H" and "G" modes.

Figure 3-14. M8A1 chemical alarm

Table 3-10. Unmasking procedures

Unmasking with M256 Series Kit	Unmasking without M256 Series Kit	MOPP Gear Exchange
This procedure takes 10 minutes. See *FM 3-4* for more information. The leader— • Checks M256 results to ensure that they are negative. • Selects one or two soldiers. The soldiers— • Move to a shady place. • Unmask for 5 minutes. • Clear and reseal their masks. The leader— • Observes the soldiers for 10 minutes. • Gives the all-clear signal if the soldiers show no symptoms. • Watches for delayed symptoms. • Has first aid available.	This procedure takes 25 minutes. See *FM 3-4* for more information. The leader— • Checks the M8 paper for liquid contamination. • Selects one or two soldiers. The soldiers— • Move to a shady place. • Break the seal of their mask, with their eyes open, for 15 seconds. • Clear and reseal their masks. The leader observes the soldiers for 10 minutes. If no symptoms occur, the soldiers— • Unmask for 5 minutes. • Clear and reseal their masks. The leader— • Observes the soldiers for 10 minutes. • Gives the all-clear signal if the soldiers show no symptoms. • Watches the soldiers for delayed symptoms. • Has first aid available	Soldiers conduct MOPP gear exchange as a two-man team. See *FM 3-5* for more information. Soldiers— • Decon their individual gear. • Prepare their mask for decon and their overgarment for removal. • Decon their mask hood. • Remove their overgarment and overshoes. • Remove their gloves. • Put on their new overgarment. • Put on their new overshoes and gloves. • Secure their hood. • Secure their gear.

Table 3-11. NBC threat, detection, and protection

Agent Type	Agent Class	Agent Effect	M8A	M256	CAM	M8	M9	Protection Required	Skin Decon	Equip Decon
AC	NP	Blood	N	Y	N	N	N	Mask only	N	N
GA	NP	Nerve	Y	Y	Y	Y	Y	Mask and overgarment	Y	N
GB	NP	Nerve	Y	Y	Y	N	N	Mask and overgarment	Y	N
GD	P	Nerve	Y	Y	Y	Yellow	Y	Mask and overgarment	Y	Y
VX	P	Nerve	N	Y	N	Green	Y	Mask and overgarment	Y	Y
HL	P	Blister	N	Y	Y	Red	Y	Mask and overgarment	Y	Y
HD	P	Blister	N	Y	Y	Red	Y	Mask and overgarment	Y	Y
CG	NP	Choking	N	N	N	N	N	Mask only	N	N

CHAPTER 4

Fire Support

The platoon gets FS in the forms of indirect fires, smoke, and obstacle emplacement. FS protects the platoon while it is executing its mission. The platoon may provide survivability to FS units. The platoon leader must—

- Keep the TOC apprised of all element locations at all times. The FSO gets information on engineer unit locations from the TOC operations section (there are no FISTs with engineers).
- Work closely with the FSO or the FIST to plan and report SCATMINEs, cover friendly obstacles with fire, and provide suppressive fires and smoke on locations of expected enemy obstacles.
- Keep track of the amount of artillery FASCAM allocated to the TF and still available for employment.
- Get the frequencies, call sign, and procedures needed to call for indirect fires for the platoon. Most call for fires will be done by adjusting fires from a target or known point. See *FM 5-34* for call-for-fire procedures. Two examples of call-for-fire procedures follow:
  - From your location, you will adjust fire by distance and direction from the TRP to the target. Give the distance in meters and the direction as you see it. If the target is to the left or right of the TRP as you see it, call LEFT distance in meters or RIGHT distance in meters. If the target is behind the TRP as you see it, call ADD distance in meters and if the target is in front of the TRP, call DROP distance in meters.
  - During breaching operations, you may have to call for smoke. Ensure that the FSO or the FIST knows the location and the duration of your smoked target. The call-for-fire procedures are similar to those discussed above.
- Determine if the maneuver commander plans to provide survivability positions for artillery assets and what the priority is for those positions.
- Ensure that obstacles are synchronized with the indirect-fire plan.

CHAPTER 5
Air Defense

The platoon leader and sergeant should understand how AD weapons can support the platoon and how the platoon can support AD units. The platoon leader should know where the AD umbrella is in relationship to his area of operations and what kind of support he can expect. The platoon can provide its own AD measures, either passive or active. Passive AD measures are used to keep soldiers and equipment undetected by enemy air assets. Active AD measures are used to engage enemy aircraft. The following are active AD measures:

- Using the football-field method for high-performance aircraft (aim 200 meters in front of the aircraft).
- Using the football-field method for helicopters (aim 50 meters in front of the helicopter).
- Aiming weapons slightly above the nose of aircraft that is approaching head on.
- Firing all weapons at the reference point
- Ensuring that the AD-weapons control status is given. Status could be—
  - Hold—weapons are fired only in self-defense or in response to a formal order.
  - Tight—weapons are fired only at aircraft positively identified as hostile.
  - Free—weapons are fired at any aircraft not positively identified as friendly.
- Ensuring that AD warning conditions are given. Conditions could be—
  - White—an air attack is not probable.
  - Yellow—an air attack is probable.
  - Red—an air attack is imminent or in progress.
- Determining if the maneuver commander plans to provide survivability positions for AD assets and what the priority is for those positions.
- Ensuring that the platoon is able to receive AD early warning. When the AD warning condition changes to red, the warning of incoming enemy aircraft and their direction is transmitted over command or operations and intelligence networks. The platoon leader should ensure that the platoon is monitoring the appropriate networks.

CHAPTER 6
Combat Service Support

The platoon's CSS is determined by its supply requirements. The platoon leader and sergeant decide what CSS the platoon needs for an operation and initiate a request. Their request is forwarded to the 1SG of either their parent engineer company or supported maneuver unit, depending on their command and support relationships. For more information on command and support relationships, see *Chapter 2*. The platoon's CSS is normally delivered in a scheduled logistical package (LOGPAC).

RESUPPLY OPERATIONS

A LOGPAC is a resupply element that is formed on the basis of a unit's logistic requirements. Normally, it consists of a petroleum, oils, and lubricants (POL) truck; an ammunition truck (Class IV/Class V supplies); and a supply truck that carries rations, repair parts, and other requested items and pulls a water trailer. LOGPACs are—

- Assembled in the BSA and then moved forward under the control of the support-platoon leader to a logistics release point (LRP).
- Picked up by the 1SG at the LRP and moved forward to a secure area behind the company's position to feed, fuel, and resupply the company.
- Returned by the 1SG to the LRP where the support-platoon leader takes control of it and moves it back to the BSA.

LOGPACs are usually distributed by one of the following methods:

- Service-station method.
- Tailgate method.
- Modification of both.

In the service-station method, the 1SG establishes the LOGPAC in a centrally located and secure site. Successively, each platoon moves to the LOGPAC site. Each squad or crew passes through the various stations before the final inspection and returns to the work sites. Normally, the stations include maintenance support, fuel (POL), food, water, mail, administrative support, medics, and other CSS assets as required by the SOP and the mission, enemy, terrain, troops, and time available (METT-T). *Figure 6-1* shows the service-station method.

The tailgate method differs from the service-station method in two major aspects (see *Figure 6-2)*. In the tailgate method, the LOGPACs—

- Are brought to each platoon's general location by the 1SG.
- Are usually not as comprehensive as the service-station LOGPACs because they are more mobile and spend more time in transit.

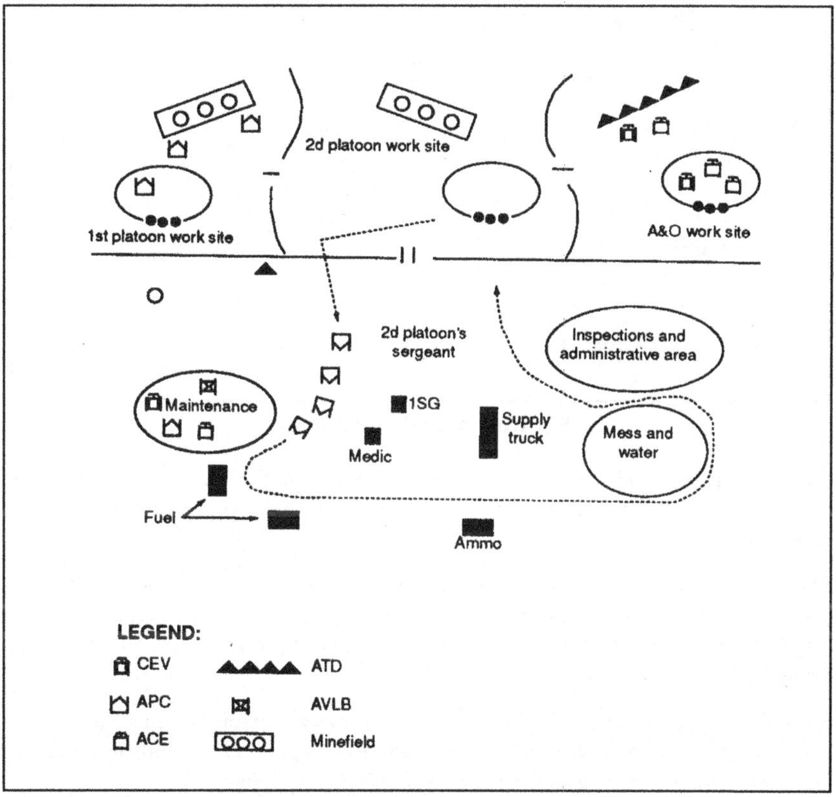

Figure 6-1. Service-station method

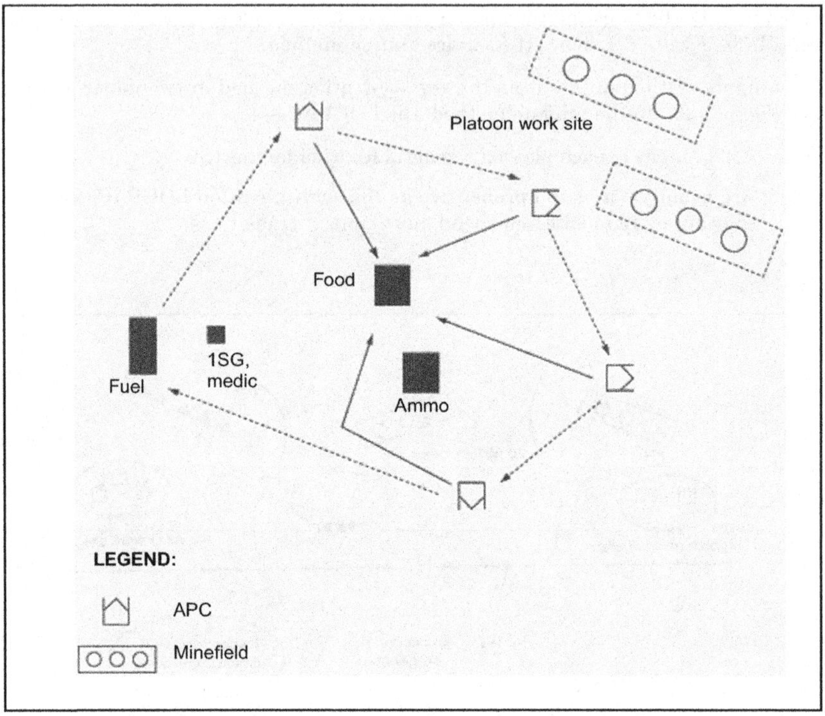

Figure 6-2. Tailgate method

The tailgate method takes more time for the 1SG but is the least interfering of the platoon's work effort. The platoon does not have to spend time moving to the LOGPAC.

Platoons often get resupplied from the maneuver company's/team's LOGPAC using the service-station or tailgate method. For security reasons, maneuver units normally provide LOGPACs to individual vehicles. Therefore, the squad rather than the platoon gets resupplied.

Even when being logistically supported by maneuver elements, platoons sometimes need engineer-specific resupply. Logistics-support responsibilities for command and support relationships are shown in *Table 6-1, page 6-4.*

Table 6-1. Command and support relationships

Relationship	Command and Task Organize	Assign Tasks	Provide Logistics Support	Reorganize as Infantry
Attached	Supported unit	Supported unit	Supported unit	Supported unit
OPCON	Supported unit	Supported unit	Parent unit	Supported unit
DS	Parent unit	Supported unit	Parent unit	Parent unit
GS	Parent unit	Parent unit	Parent unit	Parent unit

SUPPLY CLASSES

Normally, there are 10 classes of supplies; however, during defensive operations, Class IV/Class V supplies require special engineer considerations. Some supply classes require special considerations in resupplying the engineers. Supplies can either be pushed or pulled through the system. Supplies are pushed by predicted requirements and pulled by the needs of the user via a request for resupply.

CLASS I

Class I consists of subsistence and gratuitous health and welfare items. The platoon allocates Class I supplies based on unit strength. Class I supplies are normally pushed. The following are the amounts of water required for a soldier per day for—

- Drinking - 6 gallons.
- Personal hygiene - 2.5 gallons.
- Food preparation - 1.0 gallon.
- Vehicle radiators - 25 percent of the radiator's capacity.

CLASS II

Class II consists of such items as clothing, individual equipment, tentage, hand tools, administrative and housekeeping supplies, chemical-defense and decontamination items, and wet- and cold-weather contingency items. The company allocates Class II supplies based on need. It is normally pulled.

CLASS III

Class III consists of POL, including petroleum fuels, lubricants, hydraulic and insulating oils, preservatives, liquids and gases, bulk chemical products, coolants, deicer and antifreeze compounds, components and additives of petroleum and chemical products, and coal. Supply amounts are based on either offensive or defensive operations. When the TF is preparing to defend, additional Class III supplies are required for all equipment. ACEs require fuel every 8 hours when being heavily used. Additional fuel and package products, such as hydraulic fluid, should be maintained in the combat trains for this purpose. Forward fueling at the work sites should be a standard procedure. Class III supplies are usually pushed.

CLASS IV

Class IV consists of construction materials, including all fortification and barrier materials. These are items for which allowances are not prescribed. Class IV supplies are pulled.

CLASS V

Class V consists of all types of ammunition, including radiological and special weapons, bombs, explosives, mines, fuzes, detonators, pyrotechnics, missiles, rockets, propellants, and other associated items. Engineers need some unique Class V supplies (for example, cratering charges and line charges), which require special coordination with the Supply Officer (US Army) (S4) to get them into the TF's Class V system. Class V supplies are pulled.

CLASS VI

Class VI consists of all personal-demand items, such as candy, cigarettes, soap, and cameras (nonmilitary sales items). Sundry packs are also Class VI items. The platoon allocates Class VI supplies based on need. Class VI supplies are normally pulled.

CLASS VII

Class VII consists of major end items, such as launchers, tanks, mobile machine shops, vehicles, and organizational tool sets. Class VII supplies are demand driven.

CLASS VIII

Class VIII consists of medical material, including repair parts peculiar to medical equipment. Class VIII supplies are normally pulled.

CLASS IX

Class IX consists of repair parts and components, including kits, assemblies, and subassemblies (repairable and unrepairable) that are required for maintenance support of all equipment. The platoon's stock of repair parts is based on vehicle density, the authorized stockage list (ASL), the high-demand prescribed load list (PLL), and specific engineer-equipment requirements. Repair parts for items that the TF does not stock (for example, the M9 ACE or dozer) will have to come from the engineers. The platoon generates requests for repair parts by accurately completing *DA Form 2404*. When the maintenance contact team gets the form, it determines what repair parts are needed and the level of repair. The mechanics may bring the repair parts with them or the crew/operator may get them from the LOGPAC. Class V supplies are pulled.

CLASS X

Class X consists of materials to support nonmilitary programs, such as agriculture and economic development. Class V supplies are pulled.

SPECIAL CLASS IV/CLASS V SUPPLIES

These classes of supplies form the majority of the materials with which engineers construct the obstacles and fortifications required for the defense.

There are two types of Class IV/Class V loads: mission and basic. Mission loads consist of those materials required for a specific mission (for example, a standard fix minefield). Basic loads consist of those materials that the platoon carries to protect itself. The basic load can be used for missions to save time; however, it is to be replenished from the materials in the mission load. Basic loads are pulled and mission loads are pushed.

Mission loads are a TF responsibility regardless of the command and support relationships specified for engineers. The quantity of Class IV/Class V supplies (mines) normally stretches or exceeds the transportation assets of the TF. Palletized standard loads help solve the planning and distribution problem. Class IV/Class V supply points should be selected by their proximity to major obstacles.

Class IV/Class V resupply for the defense is one of the most demanding logistics operations the TF must carry out and requires all the assets that can be made available. A total cooperative effort by the TF, including engineers, is required if the defense is to be adequately resourced.

In the offense, TF planners anticipate Class IV/Class V mission loads for even a hasty defense on the objective. Prepackaging in the BSA speeds movement forward when it is needed. In addition to the Class V supplies that come forward as a part of the LOGPAC, the TF will maintain emergency Class V resupply in the combat trains.

MAINTENANCE PROCEDURES

When the TF is providing logistics support to the engineers, it will also provide organizational maintenance support to them. This causes no particular problem for common equipment. No engineer equipment repairers exist in the maneuver unit's maintenance sections lower than the brigade level except those at the engineer battalion. It may be unrealistic to expect the maneuver unit to provide all logistics support. The platoon may have to coordinate for maintenance assistance from its parent unit. When the engineers are providing logistics support, they will also provide organizational maintenance support.

Doctrinally, recovery is the responsibility of the owning unit. However, it is virtually impossible for engineers working in the TF area to recover their own vehicles. In the defense, the TF will have to provide recovery support at least back to the unit maintenance collection point (UMCP) located near the combat trains. If repairs cannot be made at the UMCP and the time and situation permit, the engineer's maintenance-support team (MST) can come forward and recover the down vehicle. The TF could provide additional recovery back to the BSA for repairs. The engineer battalion maintenance officer (BMO) or technician must ensure that each BSA has direct-support (DS) mechanics capable of repairing engineer equipment.

In the offense, the TF recovers the vehicle to the main supply route (MSR). Depending on the situation, the TF either turns vehicle recovery over to the engineers or recovery assets take the vehicle to the BSA. It must be clear in the TF SOP or order which MST has recovery responsibility for the engineer vehicles. The priority the TF commander places on the engineer assets determines the priority they are recovered in.

Engineer companies should have maintenance contact teams that are familiar with engineer heavy equipment to routinely visit work sites. The communications repairman and armorer can accompany this contact team to help anticipate problems and provide support before the mission is jeopardized.

MEDICAL SUPPORT

Medical support and evacuation must be coordinated for each operation and normally come from the nearest company/team (see *Table 6-2, page 6-8)*. The platoon's aidman or combat lifesavers should be required to coordinate with the TF's medics and ensure that the medical requirements of the platoon are planned for and met. In the event of a mass casualty incident, the platoon may have to designate a platoon vehicle as a litter vehicle.

Table 6-2. Command and support relationships and logistics support

Logistics Support	Class III	Class IV/V	Class VIII Medical	Maintenance	Emergency Resupply
Parent unit (OPCON, DS, GS)	Company LOGPAC	Basic loads are resupplied from the LOGPAC. Mission loads are supplied by the supported unit.	Parent unit	Parent unit performs maintenance on all equipment.	Parent unit
Company/team (attached)	Company/team LOGPAC	Basic loads are resupplied from the LOGPAC. Mission loads are supplied by the supported unit.	Company/team	Company/team MST performs maintenance on common equipment. Engineer MST performs maintenance on engineer-specific equipment.	Company/team
TF (attached)	TF support platoon	Basic loads are resupplied from the LOGPAC. Mission loads are supplied by the supported unit.	TF medical platoon	TF maintenance platoon performs maintenance on common equipment. Engineer MST performs maintenance on engineer-specific equipment.	TF support platoon

The platoon leader must—

- Understand the TF's evacuation and medical support plans.
- Understand the engineer company's evacuation and medical support plans.
- Know the locations of the—
  - Battalion aid station.
  - Jump aid station.
  - Personnel transfer points.
  - Casualty collection points.

CHAPTER 7
Command and Control

The platoon leader uses the C^2 processes to ensure that his platoon accomplishes its missions. This chapter provides the TTP needed to command and control the platoon and make sound tactical decisions.

COMMAND AND SUPPORT RELATIONSHIPS

Command relationships can be operational control (OPCON) or attached. Support relationships can be DS or general support (GS). Both supporting and support parties must clearly understand what each of these mean (see *Table 6-2, page 6-8*). The following should be considered when using command and support relationships:

- When OPCON is used, the maneuver element—
  - Commands and controls the platoon.
  - Has the authority to place squads OPCON to companies/teams even though it is not considered advisable to do so with a mechanized platoon. Additionally, the TF cannot attach them to the company.
- When attached is used, the maneuver element—
  - Commands and controls the platoon except for limitations imposed by the attachment order.
  - Has responsibility for the platoon's logistics support except for personnel transfer and promotion.
- When DS is used—
  - The platoon receives its missions from the maneuver element.
  - The engineer company and the engineer battalion support and command the platoon.
  - The platoon works on the tasks that the maneuver element assigns.
  - The TF does not have the authority to split the platoon up and attach squads to companies/teams.
  - Logistics support for the platoon comes from engineer higher HQ.

- When GS is used, subordinate commanders must request support from the engineer commander on a task-by-task basis, and the senior commander sets priorities and assigns tasks to the GS unit. GS is seldom used at TF level. When an engineer commander wants to retain control over the engineers, they are retained in GS of maneuver forces.

The TF always provides the materials necessary to support engineer operations in its sector for offensive and defensive missions. This is a key logistics issue that must be understood at the start of any C^2 discussion. Even though the engineers determine requirements and use the material, supplying it remains a TF responsibility regardless of the command or support relationship (see *FM 5-100*).

COMMUNICATIONS

Engineers must operate on the following frequency modulated networks:

- TF.
- Platoon internal.
- Engineer company.
- Maneuver company/team.

Engineers have a two-net capability to support their communication requirements. In the DS or GS relationship, the engineer platoon should operate the first network on the engineer command network. The second network, the platoon's internal network, should be used to control the platoon's assets. In the attached or OPCON relationship, the engineer platoon should operate the first network on the engineer command network or supported maneuver company/team command network based on the task organization. The platoon leader uses his platoon's internal network to enter other networks for coordination and situation reporting. Before the platoon leader leaves his platoon's internal network, he must first hand over control of the network to the platoon sergeant to control the platoon's assets.

The platoon submits the following reports:

- Routine reports:
  - Commander's situation report (SITREP)—identifies changes to a unit's tactical situation.
  - Sensitive-items report—gives the results of a serial-number check on all sensitive items.
  - Supply-status report—informs the commander and staff of a subordinate unit's supply status.
  - Personnel-status report—is used to maintain the accountability and status of assigned and attached personnel and generate requests for replacements.

  - Maintenance-status report—provides information on the operational status of equipment and is used to generate requests for maintenance and parts.
- Event-driven reports:
  - Spot report (SPOTREP)—identifies any known or suspected enemy activity.
  - Reconnaissance report—provides technical and tactical engineer data on a point/area target.
  - Countermobility report—gives the status of all obstacles.
  - Mobility report—gives the status and conditions of lanes, routes, bypasses, crossing sites, air lines of communication, as required, and ground lines of communication. It also includes status changes and the upgrade/downgrade of previously reported trafficability of locations/sites.
  - Survivability report—provides the detailed status of the unit's survivability level (for example, the number, type, and protective level (hull or turret) of positions).
  - Class IV/Class V special report—is used to keep track of critical engineer classes of supplies. For example, a platoon establishes a mine dump and keeps track of the types and quantities of mines available and the amounts issued.
  - Flash SITREP—is used to notify the next higher HQ when a subordinate unit becomes decisively engaged or has critical shortages that prevent mission accomplishment.
  - NBC report—is used to report NBC events.
  - Shelling report (SHELREP)—provides information on enemy indirect fire.
  - Enemy prisoner of war (EPW) report—is used to account for prisoners captured.
  - SCATMINE report—is used to notify affected units of the emplacement of SCATMINEs.
  - Meaconing, intrusion, jamming, and interference (MIJI) report—is used to notify the Intelligence Officer (US Army) (S2) when the reception of radio signals is hindered, confused, or distorted by any external source or when instructions are received from a station that cannot authenticate.
  - Splash report—is used to report downed friendly aircraft.
  - Casualty report—identifies battle and nonbattle casualties to the parent unit and alerts medical-treatment facilities of incoming casualties.
  - Casualty-feeder report and witness-statement report—provide detailed eyewitness information on a casualty. They are used to account for casualties and provide administrative information for use in notifications and awards.

ENGINEER INTEGRATION

The company commander of a mechanized engineer company normally associated with a battalion/TF is also the staff engineer and advisor to the TF commander. Likewise, in a light TF, the staff engineer advisor to the TF commander is the light engineer platoon leader. The staff engineer integrates engineers into the TF's planning process. He coordinates with other BOS to ensure engineer synchronization. During the preparation of the battle, the staff engineer ensures that soldiers and equipment are in the correct location and on time for linkups and rehearsals. He then executes the engineer mission to support the operation. The platoon is the lowest level engineer unit that can still effectively accomplish independent tasks. The platoon usually operates under the control of an engineer company; however, it can be attached to a maneuver company/team.

The platoon leader must keep both the engineer company and maneuver company/team commanders informed on critical tasks. He must continuously forward and update tactical and technical information upward and laterally by all available means.

In heavy forces, the TF normally receives an engineer company in support, and the company commander is the TF engineer. In light forces and in some special situations in heavy forces, the TF only receives an engineer platoon, and the platoon leader is the TF engineer. When the platoon leader is acting as the TF engineer, he should consider the following information:

- Orders should include when and where the TOC is to be established. If the TOC is to be moved before the platoon arrives, this should also be included.
- Orders should identify an assembly area where the platoon can be consolidated.
- The entire platoon must **not** go to the TOC.

PLATOON LEADER'S RESPONSIBILITIES

Once the platoon leader arrives at the TOC, he should—

- Check in with the S3 or company commander and report the platoon's status and location and any problems.
- Provide the status of equipment and the number of soldiers to the operations noncommissioned officer (NCO).
- Obtain the TF's and the team's signal operation instructions (SOI) and enter the maneuver force's network.

- Ensure that the communications equipment is set for the correct variable for secure operations and is compatible with the supported unit's communications equipment.
- Check with the operations sergeant for any changes to routine or required reports and SOP items.
- Be familiar with the TF SOP. **This is not the time to start studying the SOP.**
- Provide the status of obstacles being constructed.
- Check on format requirements for the periodic report. He should also check to see that TOC personnel are familiar with the reports being sent and know how to post them to their charts or overlays.
- Determine if a specific individual at the TOC has been designated as an expediter or overwatcher for engineers in the sector requiring assistance.
- Provide any terrain, route, or other information acquired from the reconnaissance unit or personnel enroute to the S2 or company/team commander.
- Obtain such information as the current enemy situation, weather and light data, and terrain details, to include topographic products, from the S2 or company/team commander.
- Provide any priority intelligence requirements (PIR) or information requirements (IR) needed for upcoming operations to the S2 or company/team commander.
- Stop at the combat trains and coordinate for classes of supplies and services when joining the maneuver force.
- Obtain frequencies and call signs for the administrative and logistics net control stations (NCSs).
- Provide a platoon roster and identify any critical shortages to the Adjutant (US Army) (S1) or operations sergeant.
- Review the TF SOP for reporting and evacuating casualties.
- Coordinate with the S4 or company/team 1SG and review the platoon's logistics status. The channel used for operational logistics for the platoon varies based on the command and support relationships.

NOTE: The platoon leader should remember that if a problem arises which will prevent the accomplishment of the mission or if normal channels are not working fast enough and logistics become an operational issue, the S3 or commander should be apprised of the problem.

As a minimum, the platoon leader must be present when the—

- Maneuver force receives the WO.
- Commander conducts reconnaissance.
- Commander makes and announces a decision and explains the intent.
- OPORD is received.

MANEUVER FORCE'S CONTROL

Normally, the platoon is kept under the engineer company's control and works for the battalion/TF commander. A major consideration in planning for the use of engineers is to maintain platoon integrity, keeping the entire platoon under the platoon leader's control. When maneuver forces are moving on two axes during mechanized offensive operations, the platoon will not place a part of the platoon on each axis. During the planning process, the engineer planner must weigh the main effort and determine which axis is critical to the mission and requires engineer assets. Placing an engineer NCO with the scout platoon when it is performing reconnaissance can help provide more specific information on the obstacles that the scouts find. Obstacle breaching can thus be preplanned. See *Appendix A* for basic formations, movement techniques, and hand-and-arm signals and *Appendix B* for checklists.

ENGINEER ATTACHMENTS

If the platoon has been augmented with AVLBs, they normally come in pairs. This is for both maintenance and tactical reasons. The AVLBs should be placed with the element that needs a gap-crossing capability. In the deliberate attack, AVLBs can be integrated individually into attack formations. Employment as a pair allows the squad leader to control them. If the AVLBs are not immediately required, they should be returned to the company's control. Curbs on the AVLB must be removed if M1 tanks with rollers are to cross the AVLB, and the roller's dog bone must be raised. If a CEV is attached to the platoon, it normally moves with the platoon.

Heavy equipment and wheeled vehicles should move with the platoon only on tactical road marches over good roads. During a movement to contact or over rougher terrain, METT-T determines where and what heavy equipment moves in the formation.

APPENDIX A

Basic Formations, Movement Techniques, and Hand-and-Arm Signals

The platoon uses a variety of mounted and dismounted formations and movement techniques to maneuver on the battlefield. This appendix gives examples of many of the basic formations, movement techniques, and hand-and-arm signals the platoon leader could use. It is not designed to be all encompassing. For more information on these topics, see *FMs 7-8, 5-34, and 71-1.*

MOUNTED MOVEMENT TECHNIQUES

The mounted engineer platoon must be proficient in moving with its maneuver counterpart. In the following paragraphs, formations, movement techniques, and actions taken during movement for the mounted engineer platoon are discussed.

WEDGE

The engineer platoon almost always maneuvers as part of another larger formation. Normally, this will either be the parent engineer company or a maneuver company/team. *Figure A-1, page A-2,* shows the formation that the platoon is most likely to use. The platoon wedge provides the most defensible formation with the easiest C^2. Ordinarily, the platoon follows a maneuver platoon if task-organized to a maneuver company/team. The platoon might lead if it is part of the engineer company formation. In either case, the wedge is the best formation to use if enemy contact is likely.

The platoon frequently receives assets from the assault and obstacle (A&O) platoon. These could include the AVLB, CEV, or armored vehicle-launched mine-clearing line charge (AVLM). These additional vehicles can strain the C^2 capability of the platoon. *Figure A-2, page A-2,* shows the technique of "tethering". The AVLB, CEV, and AVLM are tethered to individual engineer squads. Each squad leader has the responsibility of commanding and controlling an A&O platoon asset. This improves the C^2 of the platoon and provides each A&O asset with a security element as it moves across the battlefield.

Figure A-3, page A-3, shows a simple wedge formation. Note where the key leaders in the platoon are. The wedge formation uses the wingman concept. The platoon sergeant and leader are both in APCs and each has a wingman. This technique

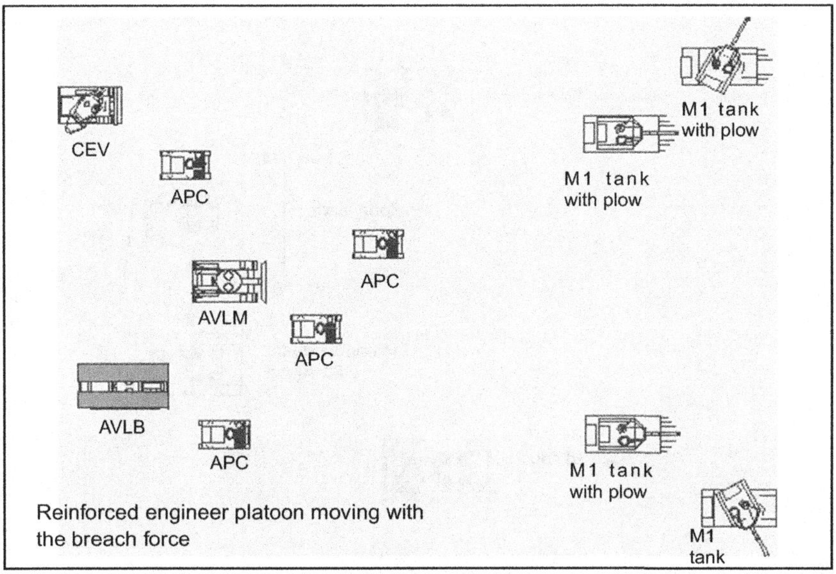

Figure A-1. Engineer platoon with the breach force

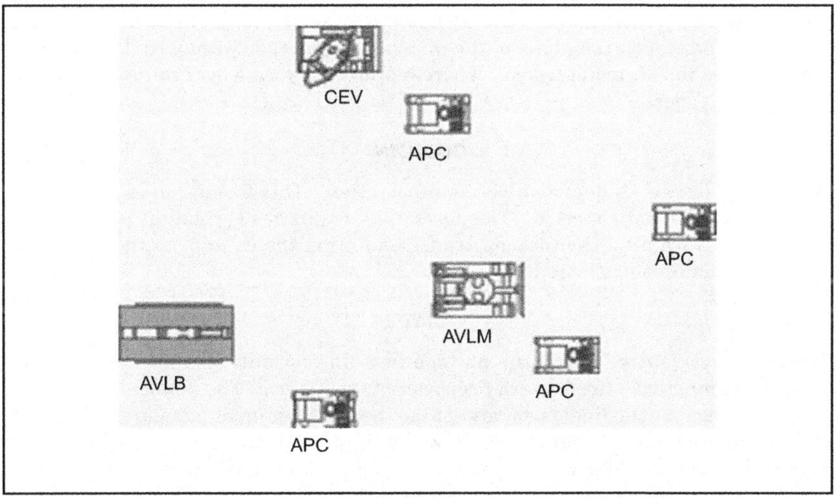

Figure A-2. Modified mounted wedge formation

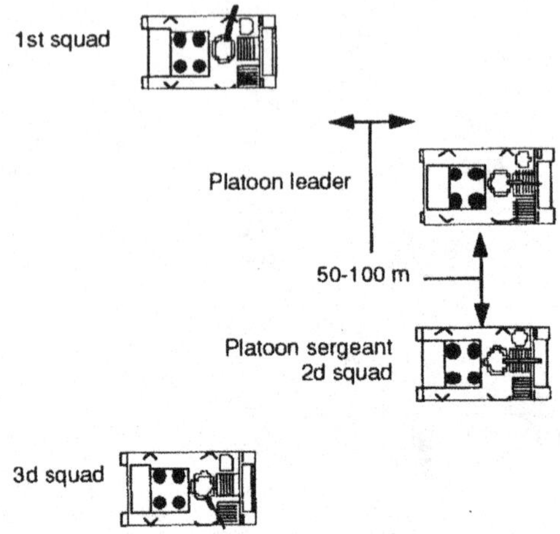

Figure A-3. Mounted wedge formation

also simplifies C^2 and movement. It has the added benefit of placing the most experienced leader in the platoon, the platoon sergeant, forward with the platoon. It also places the platoon sergeant where he can easily take over in the absence of the platoon leader.

COLUMN

Figure A-4, page A-4, depicts a column formation. This formation is used when enemy contact in not expected. This formation maximizes C^2 and the speed of the formation. Normally, the platoon transitions from the column to the wedge as enemy contact becomes more likely.

LINE

Figure A-5, page A-4, shows the platoon in a line formation. This formation is designed to maximize the forward firepower of the platoon. The platoon transitions from the wedge to the line as enemy contact becomes eminent. Engineer platoons do not generally use this formation. However, if the platoon is tasked to suppress a dismounted threat while another platoon maneuvers, this formation works well.

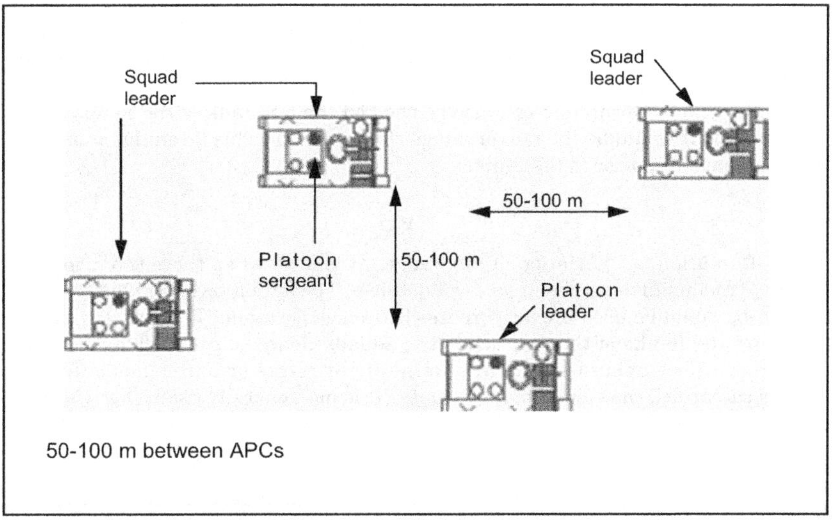

Figure A-4. Mounted column formation

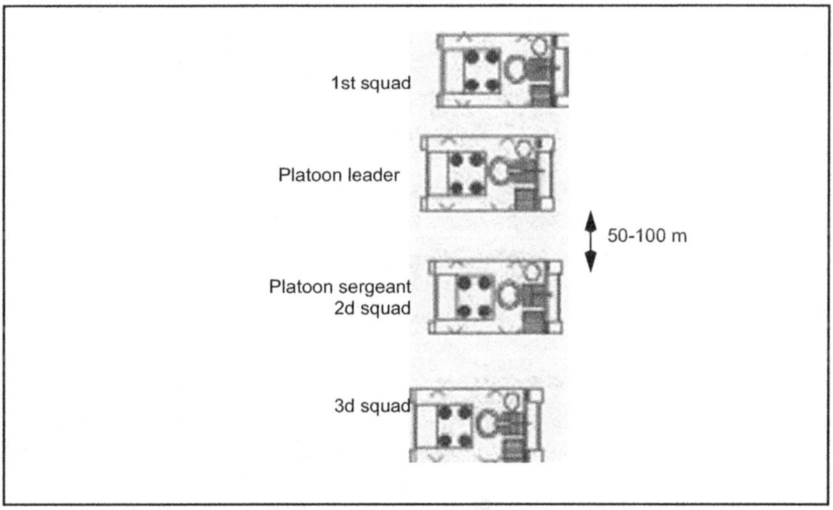

Figure A-5. Mounted line formation

ECHELON

The echelon formation is used when the platoon is on the flank of the company and the enemy threat is also from the flank. *Figure A-6* shows an echelon right formation. The echelon formation can also be used on the left flank of the company and, in that case, it would be the mirror image of *Figure A-6*. This formation maximizes the platoon's firepower to the flank.

V

The V-formation is a variation of the wedge. It is used when there is a significant threat of command-detonated mines or explosives. The formation shown in *Figure A-7, page A-6,* would be used during a route-clearance operation. It allows the platoon to secure the flanks of the road while two squads clear the route. This formation also identifies command-detonated-mine firing wires or ambushes before the enemy can attack the element on the route. It is not generally used when there is a significant enemy threat.

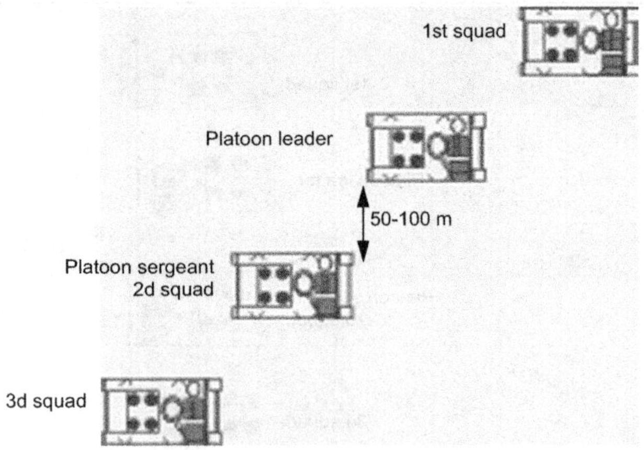

Figure A-6. Mounted echelon formation

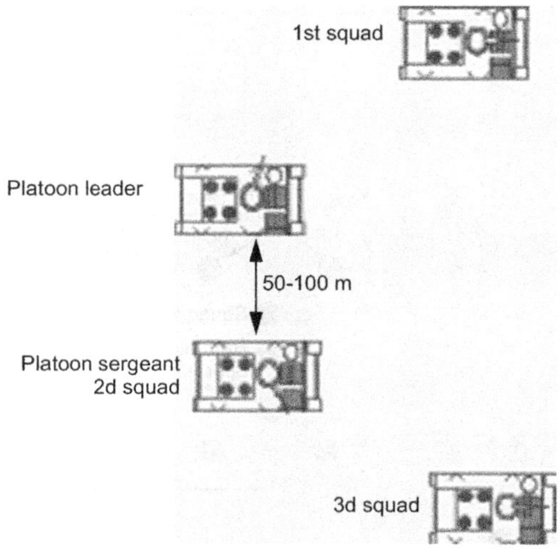

Figure A-7. Mounted V-formation

DISMOUNTED MOVEMENT TECHNIQUES

The dismounted engineer platoon must be proficient in moving with its maneuver counterpart. In the following paragraphs, formations, movement techniques, and actions taken during movement for the dismounted engineer squad and platoon are discussed.

SQUAD FORMATIONS

The squad formation is built from the fire-team wedge (see *Figure A-8)*. The interval between soldiers in the wedge formation is normally 10 meters. The wedge expands and contracts depending on the terrain. When rough terrain, poor visibility, or other factors make control of the formation difficult, the squad uses the modified wedge. In this formation, the normal interval between soldiers is reduced so that all team members can still see their team leader and each team leader can see their squad leader. The sides of the wedge can contract to the point where the wedge resembles a single file. When moving in less rugged terrain, where control is easier, soldiers expand the formation or resume their original positions.

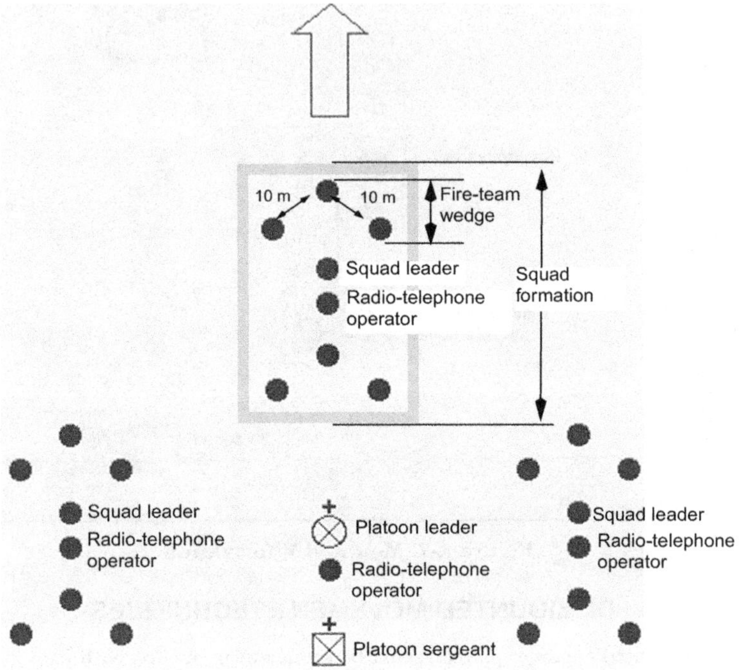

NOTE: Unlabeled positions are dismounted soldiers.

Figure A-8. Dismounted-platoon wedge formation

PLATOON FORMATIONS

Platoon formations are built from multiple squad formations. These formations include the platoon column, line, V, and wedge. The platoon leader selects the best formation based on his METT-T analysis. *Table A-1, page A-8*, shows a comparison of each formation and its use.

Table A-1. Comparison of platoon formations

Dismounted Movement Formations	Movement Use	Movement Characteristics			
		Control	Flexibility	Fire Capability/ Restriction	March Rate
Column	When the platoon performs primary movement formations	Good for maneuver (fire and movement)	Provides good dispersion laterally and in depth	Allows limited firepower to the front and rear, high volume to the flank	Good
Line	When the enemy situation is unknown and the leader wants all soldiers forward for maximum firepower to the front	Difficult	Is minimal	Allows maximum firepower to the front, little to the flanks and rear	Slow
V	When the enemy situation is vague, but contact is expected from the front	Difficult	Provides two squads up front for immediate firepower and one squad to the rear for movement upon contact from the flank	Allows an immediate heavy volume of firepower to the front or flanks	Slow
Wedge	When the enemy situation is vague, but contact is not expected	Difficult but better than the platoon V and platoon line	Enables the leader to make contact with the smallest element and still have two squads to maneuver	Provides a heavy volume of firepower to the front or flanks	Slow but faster than the platoon V
File	When visibility is poor due to terrain or light	Easiest	Is the most difficult formation from which to maneuver	Allows immediate fires to the flanks; prevents focused fires to the front and rear	Fastest

Wedge

The dismounted wedge is used when the enemy situation is vague and enemy contact is likely (see *Figure A-8, page A-7)*. This formation allows a large volume of fire around the formation. Generally, at least one squad or element will be free to maneuver from the wedge after contact.

Column

The column formation is the platoon's primary movement formation (see *Figure A-9, page A-10*). It provides good dispersion both laterally and in depth and simplifies control. The lead squad is the base squad for fire control.

Line

The line formation allows the delivery of maximum fire to the front but little fire to the flanks (see *Figure A-10, page A-11)*. This formation is hard to control and does not lend itself well to rapid movement. It is the basic platoon assault formation during an attack.

V

The V-formation has two squads up front to provide a heavy volume of fire on contact (see *Figure A-11, page A-12)*. It also has one squad in the rear that can either overwatch or trail the other squads. This formation is hard to control, and movement is slow.

File

The file formation gives maximum control to leaders and is used for speed during movement (see *Figure A-12, page A-13)*.

Traveling

The dismounted engineer platoon supporting a maneuver TF normally travels as part of one of the larger maneuver elements. When breaching assets are needed forward with the lead company, the engineer platoon travels behind the lead platoon for security (see *Figure A-13, page A-14)*. If each maneuver company needs a breaching capability, the dismounted engineer platoon can be broken into squad-sized elements, where each squad travels with a maneuver company. In this case, the platoon HQ travels with either the battalion tactical operations center or the main effort for C^2.

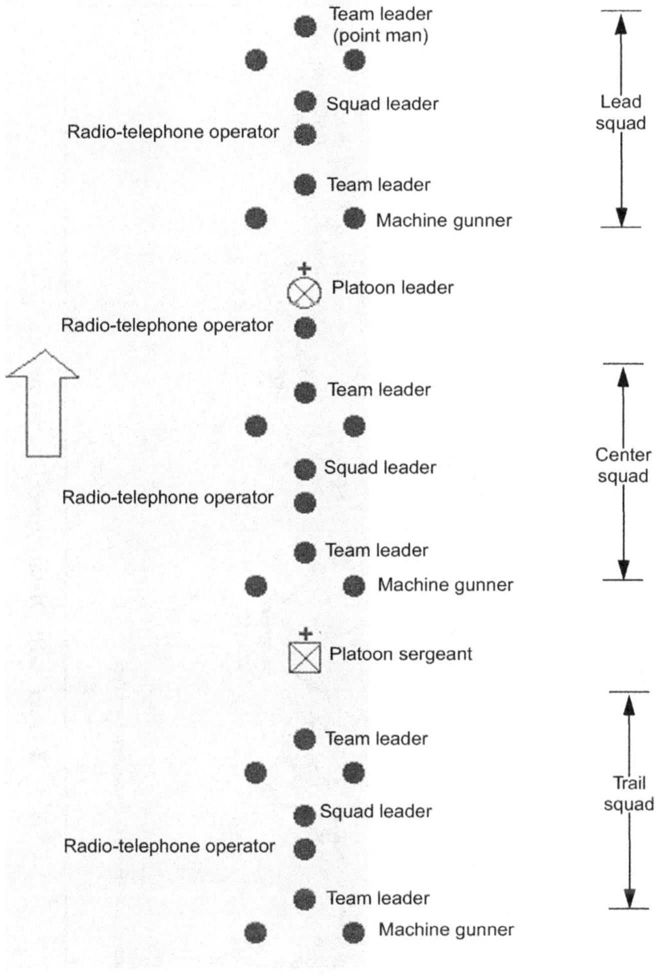

NOTE: Unlabeled positions are dismounted soldiers.

Figure A-9. Dismounted column formation

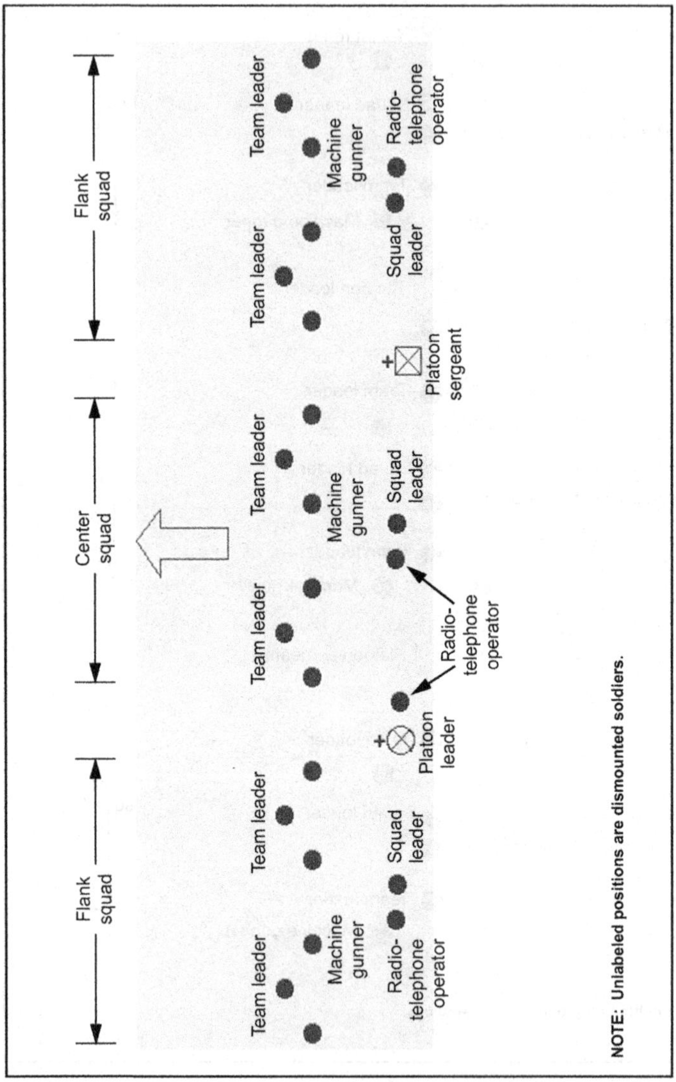

Figure A-10. Dismounted line formation

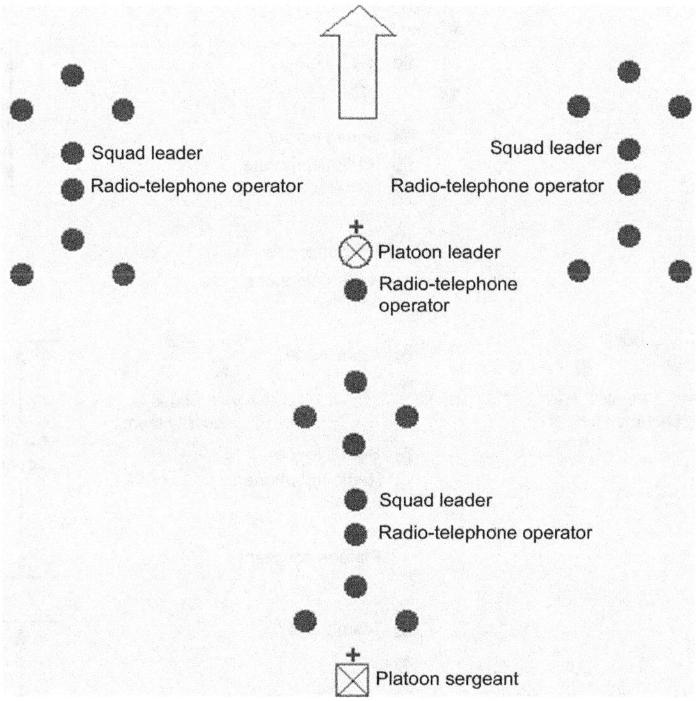

NOTE: Unlabeled positions are dismounted soldiers.

Figure A-11. Dismounted V-formation

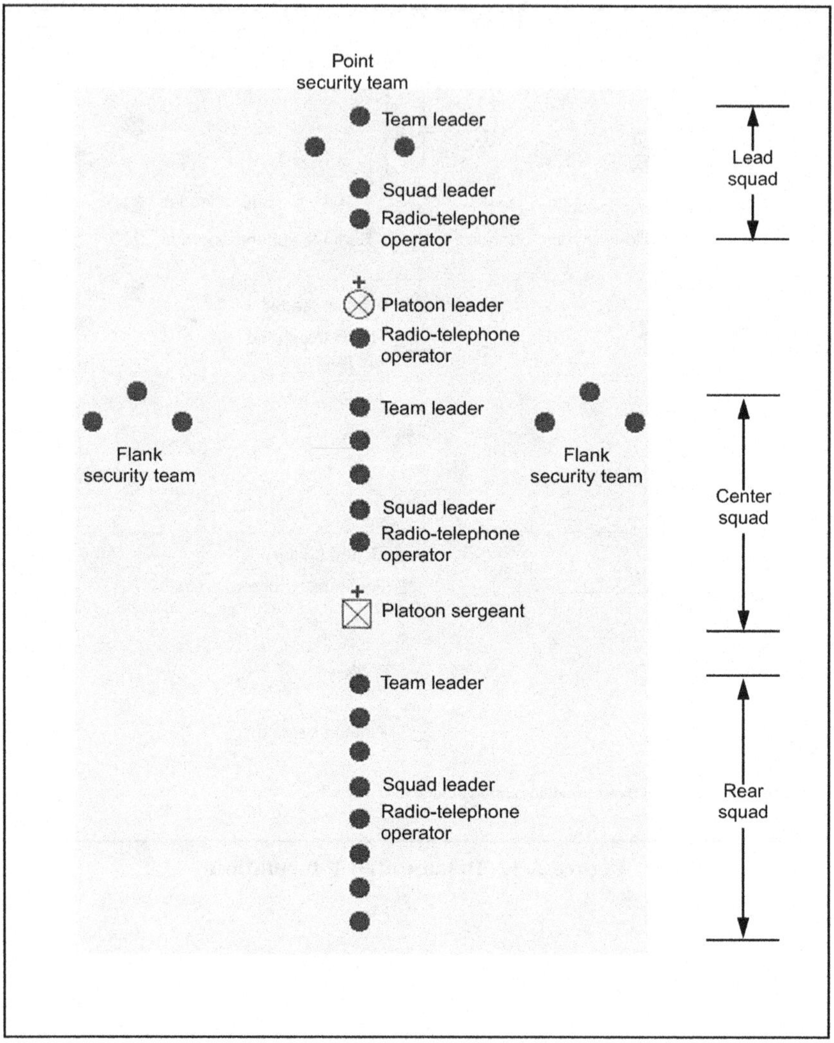

Figure A-12. Dismounted file formation

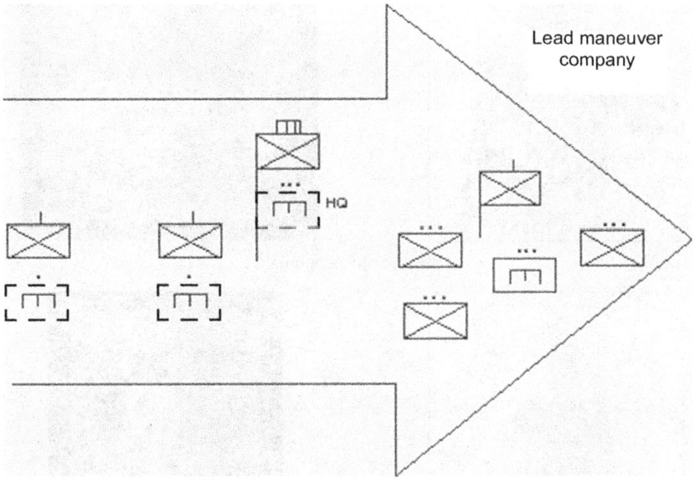

NOTE: The engineer platoon traveling with a TF could travel as part of the lead infantry company to provide immediate breaching capability, or the platoon could be broken down into squad-sized elements to travel with each maneuver company.

Figure A-13. Dismounted traveling formation

HAND-AND-ARM SIGNALS

During many operations, the platoon has to use hand-and-arm signals for C^2. *Figures A-14 and A-15, pages A-15 through A-20,* show many of the command hand-and-arm signals used when the platoon is mounted or dismounted. These signals are normally used when either radio listening silence is in effect or stealth is needed.

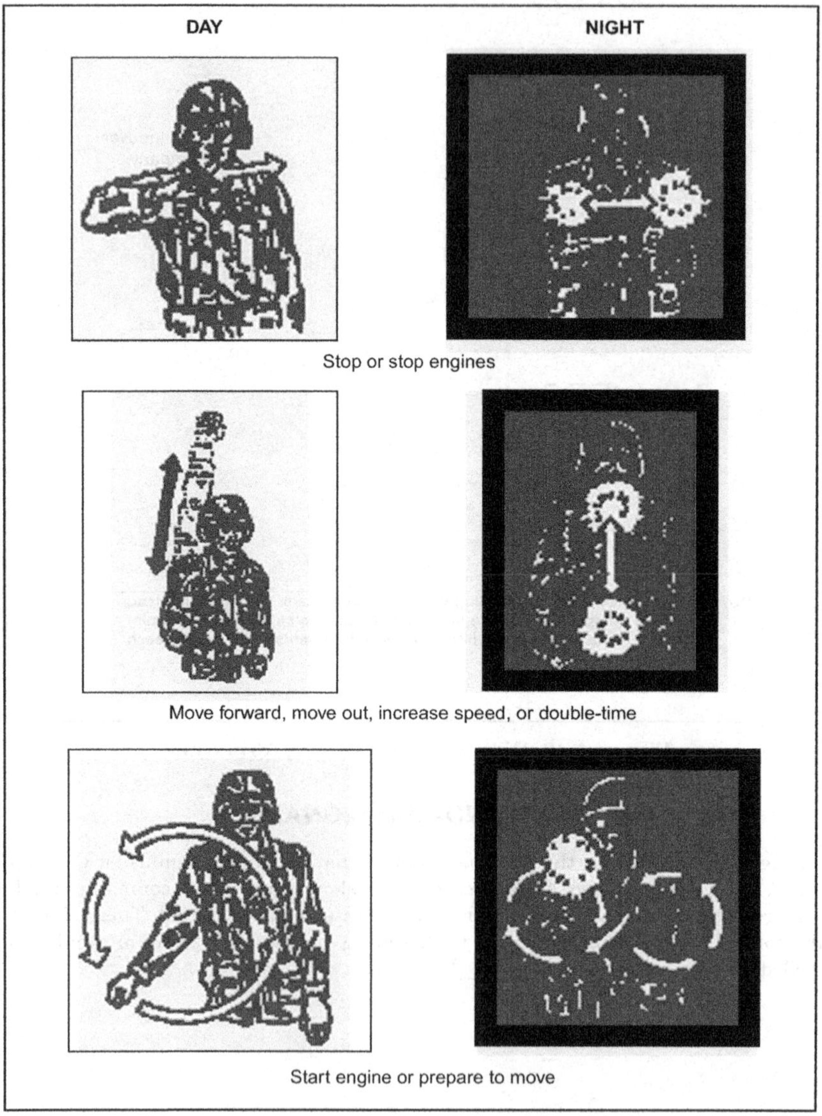

Figure A-14. Hand-and-arm signals

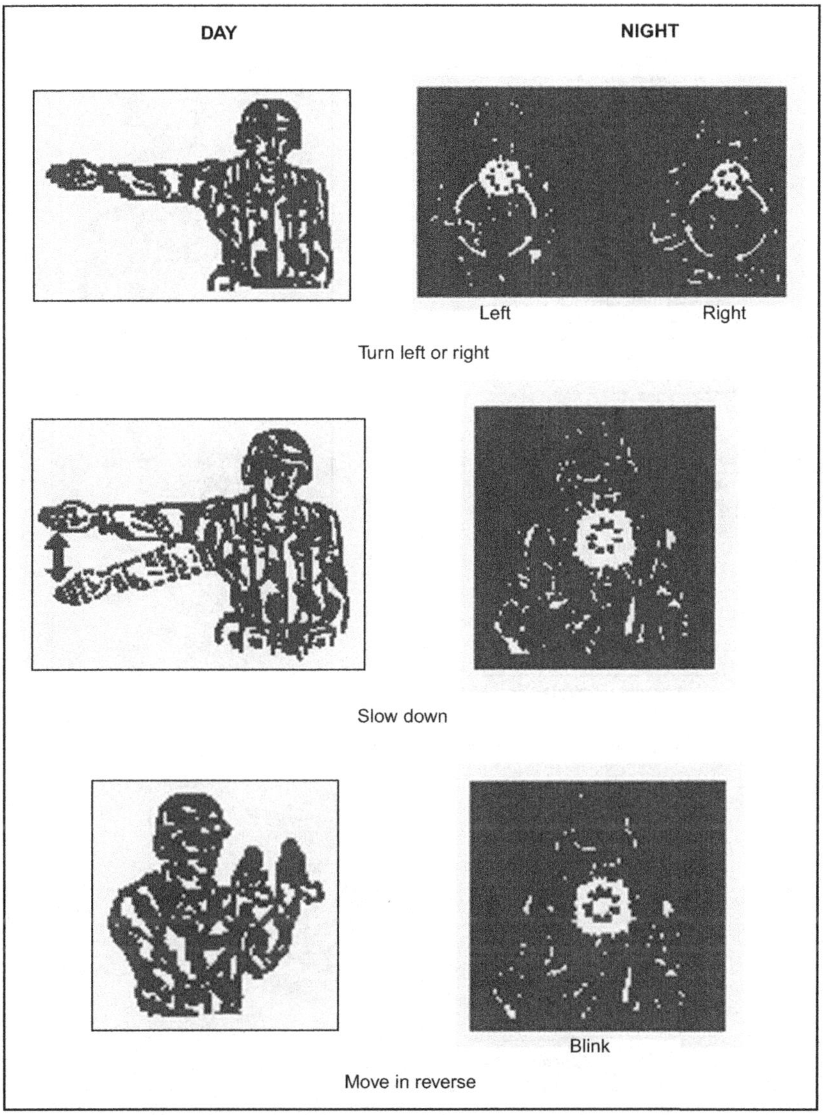

Figure A-14. Hand-and-arm signals (continued)

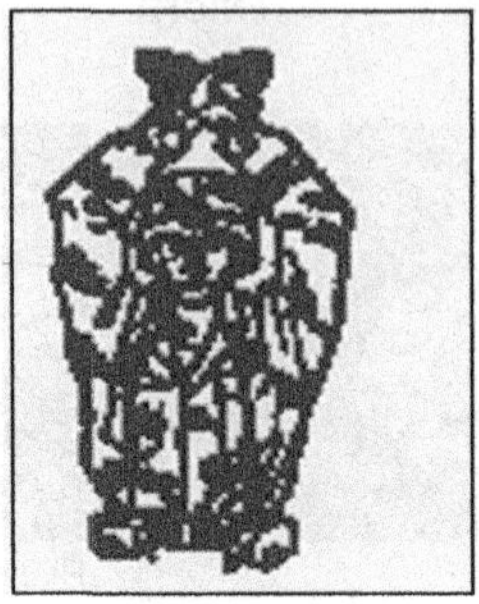
Disregard previous command

Attention

I do not understand

I am ready

Close the distance between vehicles and stop

Figure A-14. Hand-and-arm signals (continued)

Stop

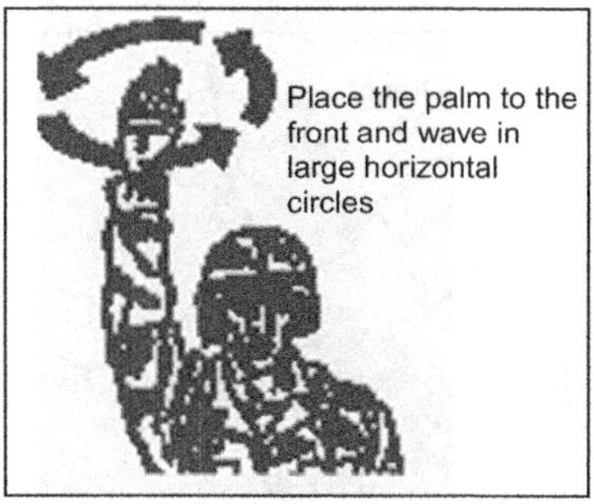

Assemble

Enemy is in sight

Commence firing

Cover our move

Cease firing

Figure A-14. Hand-and-arm signals (continued)

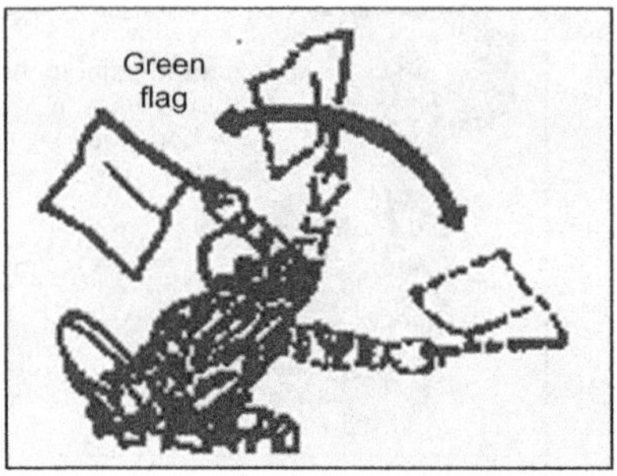

Move out

Enemy is in sight

Form a line

Figure A-14. Hand-and-arm signals (continued)

Mounted Movement Formations	Hand-and-Arm Signals	Formation Examples
V		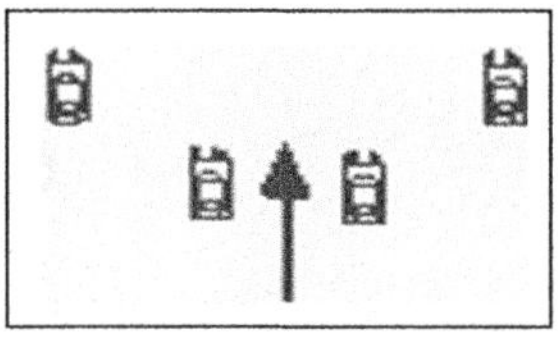
Echelon right (left)		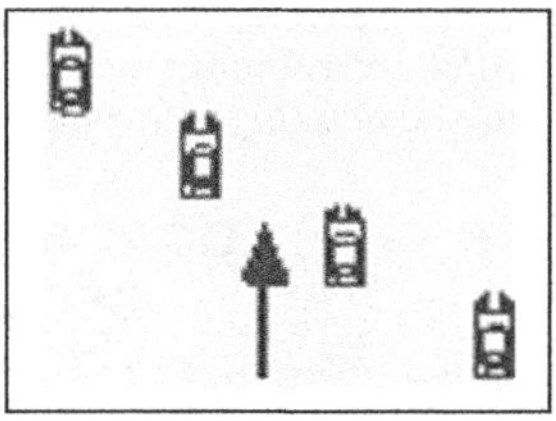
Wedge		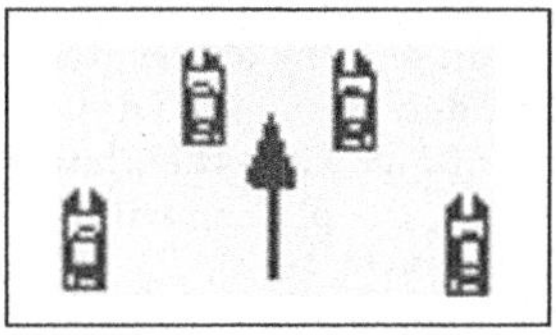
Line		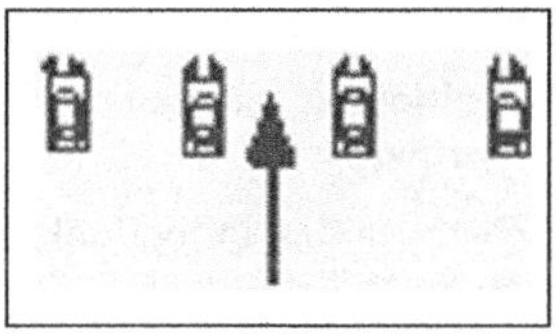
Column	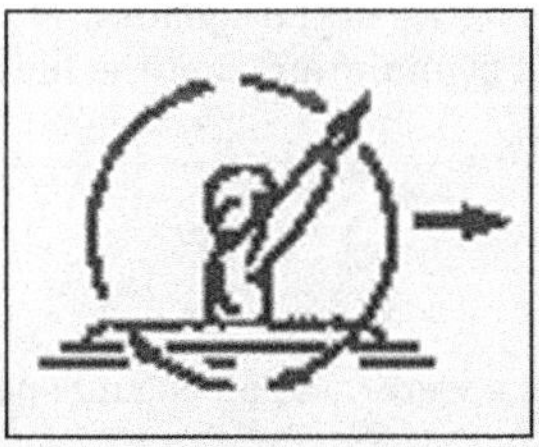	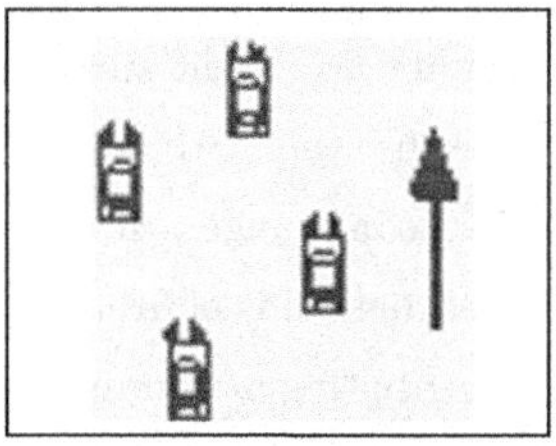

Figure A-15. Mounted movement formations and hand-and-arm signals

APPENDIX B

Checklists

This appendix gives several checklists that the platoon and its leaders will find useful. These checklists are not all encompassing and should be used to augment local SOPs, not replace them. Where units do not have standardized checklists, these may be helpful in developing local unit SOPs.

LOCAL SECURITY AND JOB-SITE SECURITY CHECKLISTS

When the platoon halts, the platoon leader establishes and maintains local security. He assigns each squad a sector of the perimeter to ensure mutual support and to cover all gaps by observation and fire. The platoon leader designates OPs and squad leaders designate OP personnel. OPs communicate with the platoon's command post and warn the platoon of the approaching enemy before it is attacked. Frequently, the platoon will be in a tactical halt while waiting for an operation to start.

When the platoon is halted, the platoon leader establishes a priority of work, to include—

- Positioning crew-served weapons, protective obstacles, and chemical alarms.
- Constructing individual and crew-served fighting positions, if the halt is for an extended period.
- Setting up wire communications between the squads and the platoon's command post. Radio silence should be maintained at the halts for OPSEC.
- Coordinating with adjacent units.
- Preparing range cards.
- Clearing fields of fire.
- Distributing ammunition, rations, water, supplies, and special equipment.
- Conducting PMCS on weapons, vehicles, and equipment.
- Inspecting the platoon's members and equipment.

- Rehearsing critical aspects for the upcoming mission.
- Test firing small-arms weapons, if the tactical situation permits.
- Conducting personal hygiene and field sanitation.
- Instituting a rest plan.
- Completing the work priorities as time permits.

Job-site-security actions are similar to those actions that are taken at a halt. However, job-site security is specifically designed to ensure that the platoon is able to work without being harassed.

Typically, the platoon will use this checklist when constructing obstacles and field fortifications. Before moving to the job site, the platoon leader informs everyone of the warning signals, code words, and pyrotechnics signals to be used. Upon arrival at the job site, the platoon—

- Occupies a job-site overwatch position.
- Dispatches reconnaissance/minesweeping/NBC teams to clear the job site.
- Establishes local security.
- Establishes escape routes; identifies possible enemy avenues of approach; and places OPs, chemical alarms, and crew-served- and antitank-weapons positions to cover enemy avenues of approach.
- Divides the job site into defensible sectors and assigns them to the squads.
- Establishes communications with the engineer company.
- Starts work.

PRECOMBAT CHECKLISTS

Each individual should have—

- Load-bearing equipment.
- A chemical protective mask with an M256 chemical-detection kit, an M13 decontamination kit, a waterproof bag for the mask, M8 chemical-detection paper, and three chemical-antidote injectors.
- A kevlar helmet with a cover and a band.
- An individual weapon with magazines and the zero posted in the pistol grip and a cleaning kit.
- A ruck sack and duffle bags packed with military clothing and table of allowance (TA)-50 equipment.

- Wire gauntlets.
- A fragmentation vest.

Each leader should have—

- A map and a compass.
- Signal operating instructions.
- His unit's SOP.
- *FMs 5-34 and 5-10* in their uniform cargo pockets.
- *FMs 20-32, 90-13-1, and 90-7* in their ruck sack.
- Other FMs, as required or needed.
- A watch and a solar-powered calculator.
- A notebook, pens, and pencils.
- Binoculars.
- Night-vision goggles.
- Casualty cards.
- A laminated range card.
- *DA Forms 1355 and 1355-1-R.*

All vehicles should have—

- Material, tied down and loaded according to the load plan.
- Fuel tanks, full.
- Preventive maintenance checks and services (PMCS), completed and dispatched.
- Packaged POL products, stowed.
- Water and extra fuel cans, full.
- Meals, ready-to-eat (MREs), stored.
- Basic-issue items (BIIs) and additional authorized list (ALL) items, inventoried, cleaned, and stowed.
- Spare parts and tires or track blocks, stowed.
- First-aid kits, inspected and stowed.
- A vehicle log book, the vehicle's TM, and *DA Forms 2404,* stowed.

- A camouflage screen system, stowed.
- Chemical-detection equipment, mounted.
- A tow cable with locking pins, stowed.
- VS17 marking panels for emergency marking.
- Ammunition evenly distributed and demolitions properly segregated and prepared for use.

UNIFORM-, DUFFLE-BAG-, AND RUCK-SACK-PACKING CHECKLISTS

A field uniform should include—

- A kevlar helmet with a cover and a band.
- A fragmentation vest.
- A pistol belt with ammunition pouches.
- A canteen with a cover and a cup, on the right hip.
- Ear plugs.
- A flashlight, on the right suspender strap (optional).
- A holster with a lanyard (if armed with a pistol), on the right side of the pistol belt.
- A bayonet, on the left side of the pistol belt.
- A compass, on the left side of the pistol belt.
- An M17A1 protective mask, on the left hip, and an M25 mask under the left arm.
- A battle dress uniform (BDU) with black boots.
- A field jacket.
- A chemical-decontamination kit connected to the protective mask.
- A nerve-agent antidote.
- A first-aid pouch, on the left suspender strap.

One duffle bag should contain—

- Three sets of BDUs.
- Three pairs of underwear.

- Three pairs of wool socks.
- Three T-shirts.
- One towel.
- Two laundry bags.
- One waterproof bag.
- One pair of boots with shoe polish, a brush, and a sewing kit.

The additional duffle bag should contain—

- Five tent pegs (10 for officers).
- Three tent poles (6 for officers).
- One tent rope (2 for officers).
- One sleeping bag.
- One pair of trouser suspenders.
- Two pairs of coveralls.
- One field jacket with a liner.
- One shelter half (2 for officers).
- One wool blanket.
- One pair of field pants.
- One pair of field pants with a liner.
- One cold-weather mask.
- One pair of extreme cold-weather boots.

The ruck sack should contain—

- Wet-weather pants and a parka.
- Overshoes.
- A weapons cleaning kit.
- Rations.
- A waterproof bag.
- A mission-oriented protective posture (MOPP) suit with gloves.
- Cold-weather mittens, winter use only.
- A parka with a liner and a hood, winter use only.

- A wool sweater, winter use only.
- A pile cap, winter use only.
- A scarf, winter use only.
- A towel and toilet articles.
- A pair of underwear.
- A T-shirt.
- Two pairs of wool socks.
- A BDU hat.
- An entrenching tool.

APPENDIX C

Breaching Case Studies

FM 90-13-1 and *Chapter 3* provide doctrinal overviews for breaching, breaching tenets, and breaching theory. The following case studies are designed to provide the platoon with instructive examples of dismounted and mounted breaching. Either case study could be applied to many different field applications. For example, the dismounted case study could be used to examine covert breaching, assault breaching, or deliberate breaching by light engineers. Likewise, the mounted case study could apply to both deliberate and in-stride breaching. These case studies are not designed to be prescriptive. They are designed to illustrate the thought process that the platoon leader must go through to successfully breach.

All breaching operations follow a similar pattern. They differ in technique. During breaching, the—

- Platoon leader determines the obstacle location and its composition.
- Platoon leader selects the breach site and lane locations.
- Platoon leader orders the breach to start, if the platoon is the primary reduction force.
- Squad prepares to breach, if the platoon is the backup reduction force.
- Squad breaches the obstacle.
- Squad marks the lane.
- Platoon leader reports the lane location to the company commander.

DISMOUNTED CASE STUDY

The company commander orders the dismounted platoon leader that is attached to him to breach an obstacle. The company is assaulting an enemy objective. The platoon leader moves to a location where he can see the obstacle and determines that it is a concertina fence and antitank minefield. He orders a squad forward to conduct reconnaissance and ascertain the best lane locations. The squad leader reports the locations and marks them with orange tape. The squad has determined that there are no antipersonnel mines in the minefield. The squad leader reports this to the platoon leader (see *Figure C-1, page C-2)*.

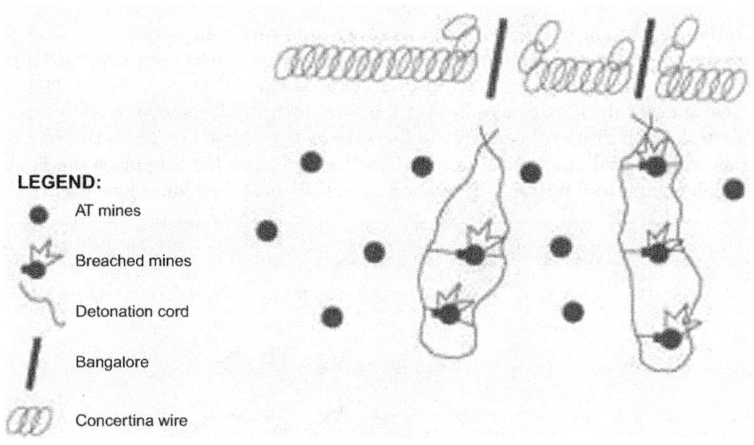

1 – The platoon leader determines the obstacle location and its composition.
2 – The platoon leader selects the breach site and lane locations.
3 – The squad breaches the concertina fence using cutters or Bangalores.
4 – The squad breaches the minefield using hand-emplaced explosives.
5 – The squad marks the lanes for the assault force.
6 – The platoon leader reports to the company commander that the lane is open.

Figure C-1. Dismounted case study

The platoon leader then orders the forward squad to remain at the obstacle as guides. He orders his second squad to come forward with two Bangalore sets. The first squad guides the second squad to the lane locations. Both squads breach the lanes in the concertina using the Bangalores and then move through the minefield placing charges to destroy antitank mines in the proposed lanes. The squads move to safety and initiate the explosives destroying the mines. Both squads provide security while the third squad comes forward to mark the lanes and guide the infantry company through the breach. The platoon leader keeps the third squad in reserve in case one of the squads fail or require additional help. However, in this scenario that was not necessary. The platoon leader observes the lanes and reports their completion to the infantry company commander who then starts his assault. The platoon leader then switches to the engineer company frequency and gives his company commander a situation report in reference to the lanes.

MOUNTED CASE STUDY

The engineer platoon is fighting as part of an engineer company/team. The company/team, which includes the engineer line platoon, the A&O platoon, and a tank platoon equipped with track-width mine plows, is the TF breach force. The company commander decides to use the tank plows, initially for speed and security, to create the breach site and secure the far side of the obstacle. Then the engineer platoon widens and marks the lanes. The A&O platoon leader places the far-lane recognition signals and initially orients the TF on the breached lanes (see *Figure C-2)*.

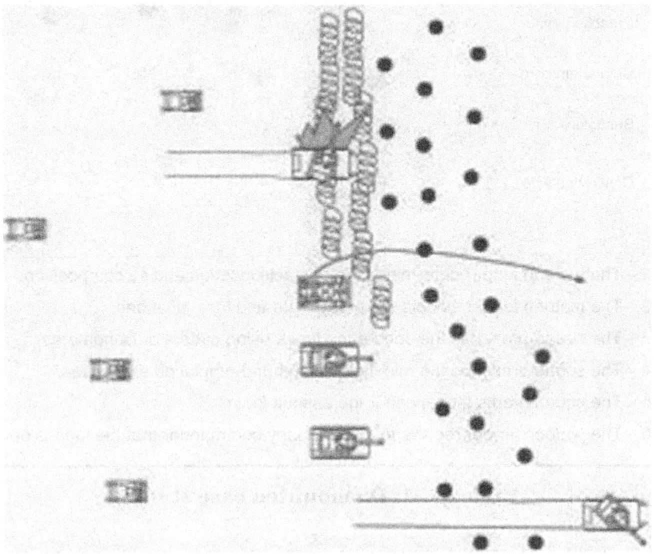

1 – The platoon leader determines the obstacle location and its composition.

2 – The platoon leader selects the breach site and lane locations.

3 – The tank crew breaches the obstacle with the tank plows.

4 – The platoon leader orders the AVLM forward to breach with the MICLIC when the enemy destroys a tank plow.

5 – The squad marks the lanes for the assault force.

6 – The platoon leader reports to the company commander that the lane is open.

Figure C-2. Mounted case study

In this scenario, the engineer company commander orders the engineer platoon forward to identify the obstacle location and its composition and potential lane locations. The platoon leader orders a dismounted element forward to conduct his reconnaissance. He reports his reconnaissance to the company commander. He states that the obstacle is a mixed minefield with antitank and antipersonnel mines, and a partial, multiple row of concertina fence is forward of the minefield. The platoon leader also reports two potential breach sites with an alternate site between them.

The engineer company commander orders the tank platoon to breach two lanes, based on the engineer platoon leaders' reconnaissance. He orders the engineer platoon to be prepared to back up the tank platoon. The northern tank equipped with a mine plow is destroyed by direct fire as it starts to breach. The engineer platoon leader observes this and orders his AVLM forward to breach with the MICLIC. The AVLM and the other tank plow successfully breach the obstacle.

The platoon leader orders two squads forward to mark each lane. These squads widen the lanes, per the initial plan, with hand-emplaced explosives after the assault force passes through. The platoon leader keeps one squad in reserve, with the AVLM, to counter any enemy attempts to close the lanes with artillery-delivered mines. He reports to the engineer company commander that the lanes are marked.

Glossary

&	and
+	plus
#	number
%	percent
1SG	first sergeant
1st	first
2d	second
3d	third
A&O	assault and obstacle
AA	axis of advance
AAR	after-action review
ABF	attack by fire
AC	hydrogen cyanide
ACE	armored combat earthmover, M9
AD	air defense
ADA	air-defense artillery
ADAM	area denial artillery munition
AHD	antihandling device
ALL	additional authorized list
ammo	ammunition

AP	antipersonnel
APC	armored personnel carrier
app	appendix
arty	artillery
ASL	authorized stockage list
aslt	assault
AT	antitank
AT-3, AT-5, AT-8	antitank missiles
ATD	antitank ditch
attn	attention
AVLB	armored vehicle-launched bridge
AVLM	armored vehicle-launched MICLIC
BDU	battle dress uniform
BII	basic-issue items
BMO	battalion maintenance officer
BMP	Soviet tracked amphibious infantry combat vehicle
BMP-1, BMP-2	Soviet infantry fighting vehicles
BOS	Battlefield Operating Systems
BSA	brigade support area
BT	blade team
C^2	command and control
CA	combined arms
CAM	chemical agent monitor
CATK	counterattack
CDR	commander

CEV	combat engineer vehicle
CG	phosgene
℄	centerline
COA	course of action
CP	checkpoint
CSS	combat service support
cu	cubic
DA	Department of the Army
decon	decontaminate/decontamination
DP	dispersing point
DS	direct support
DSA	division support area
EA	engagement area
EPW	enemy prisoner of war
FASCAM	family of scatterable mines
FEBA	forward edge of the battle area
FIST	fire-support team
FLOT	forward line of own troops
FM	field manual
FMFM	fleet Marine force manual
FS	fire support
FSO	fire-support officer
ft	foot (feet)

GA	tabun
GB	sarin
GD	soman
GEMSS	Ground-Emplaced Mine-Scattering System
GS	general support
HD	distilled mustard
HDP	hull-down position
HEMTT	heavy expanded mobility tactical truck
HL	mustard lewisite
HMMWV	high mobility multipurpose wheeled vehicle
HQ	headquarters
hr	hour(s)
IOE	irregular outer edge
IPB	intelligence preparation of the battlefield
IR	intelligence requirements
km	kilometer(s)
lb	pound(s)
LD	line of departure
LOGPAC	logistical package
LRA	local reproduction authorized
LRP	logistics release point
m	meter(s)

M-S	Miznay-Schardin
METT-T	mission, enemy, terrain, troops, and time available
MICLIC	mine-clearing line charge
MIJI	meaconing, intrusion, jamming, and interference
min	minute(s)
mm	millimeter(s)
MOPMS	Modular Pack Mine System
MOPP	mission-oriented protective posture
MP	military police
MRB	motorized rifle battalion
MRC	motorized rifle company
MRE	meal, ready-to-eat
MRP	motorized rifle platoon
MSR	main supply route
MST	maintenance support team
N	no
NA	not applicable
NAI	named area of interest
NBC	nuclear, biological, chemical
NCO	noncommissioned officer
NCS	net control station
No	number
NP	nonpersistent
OBSTINTEL	obstacle intelligence
OCOKA	observation and fields of fire, cover and concealment, obstacles, key terrain, and avenues of approach

OP	observation post
OPCON	operational control
OPLAN	operation plan
OPORD	operation order
OPSEC	operations security
ORP	objective rally point
oz	ounce(s)
P	persistent
PCC	precombat check
PCI	precombat inspection
PDM	pursuit-deterrent munition
PIR	priority intelligence requirements
PL	phase line
PLD	probable line of deployment
PLL	prescribed load list
plt	platoon
PMCS	preventive maintenance checks and services
POL	petroleum, oils, and lubricants
psn	position
R&S	reconnaissance and surveillance
RAAM	remote antiarmor mine
recon	reconnaissance
reqd	required
S&P	stake and platform
S1	Adjutant (US Army)

S2	Intelligence Officer (US Army)
S3	Operations and Training Officer (US Army)
S4	Supply Officer (US Army)
SCATMINE	scatterable mine
sec	second(s)
SEE	small emplacement excavator
SHELREP	shelling report
SITREP	situation report
SOI	signal operation instructions
SOP	standing operating procedure
SOSR	suppress, obscure, secure, reduce
SPOTREP	spot report
T/C/S	tasks, conditions, and standards
T72/80	Soviet tank
TA	table of allowance
TC	tank commander
TCP	traffic-control post
TDP	turret-down position
TF	task force
TLP	troop-leading procedures
TOC	tactical operations center
TRP	target reference point
TTP	tactics, techniques, and procedures
UMCP	unit maintenance collection point
US	United States

USAF	United States Air Force
VX	VX
w/	with
WO	warning order
WP	white phosphorous
wt	weight
XO	executive officer
Y	yes

Index

References

SOURCES USED

These are the sources quoted or paraphrased in this publication.

Army Publications

FM 3-4. *NBC Protection* (FMFM 11-9). 29 May 1992.

FM 3-5. *NBC Decontamination* (FMFM 11-10). 17 November 1993.

FM 5-34. *Engineer Field Data.* 14 September 1987.

FM 5-100. *Engineer Combat Operations.* 22 November 1988.

FM 5-103. *Survivability.* 10 June 1985.

FM 7-8. *Infantry Rifle Platoon and Squad.* 22 April 1992.

FM 20-32. *Mine / Countermine Operations.* 30 September 1992.

FM 71-100. *Division Operations.* 16 June 1990.

FM 90-7. *Combined Arms Obstacle Integration.* 29 September 1994.

FM 90-13-1. *Combined Arms Breaching Operations.* 28 February 1991.

FM 21-60. *Visual Signals.* 30 September 1987.

DOCUMENTS NEEDED

These documents must be available to the intended users of this publication.

Department of the Army Forms

DA Form 1155. *Witness Statement on Individual.* 1 June 1966.

DA Form 1156. *Casualty Feeder Report.* 1 June 1966.

DA Form 1355. *Minefield Record.* March 1987.

DA Form 1355-1-R. *Hasty Protective Minefield Record (LRA)*. July 1975.

DA Form 2028. *Recommended Changes to Publications and Blank Forms*. 1 February 1974.

DA Form 2404. *Equipment Inspection and Maintenance Worksheet*. 1 April 1979.

Made in the USA
Monee, IL
04 November 2026

67378034R00144